Plant Protoplasts

A BIOTECHNOLOGICAL TOOL FOR PLANT IMPROVEMENT

Teresa Bengochea

and

John H. Dodds

International Potato Center
Lima
Peru

LONDON NEW YORK

Chapman and Hall

First published in 1986 by
Chapman and Hall Ltd
11 New Fetter Lane, London EC4P 4EE
Published in the USA by
Chapman and Hall
29 West 35th Street, New York NY 10001

Printed in Great Britain at the University Press,
Cambridge

ISBN 0 412 26890 6 (Hb)
0 412 26640 7 (Pb)

British Library Cataloguing in Publication Data

Bengochea, Teresa
Plant protoplasts: a biotechnological tool for plant improvement.— (Outline studies in biology)
1. Protoplasts 2. Plant cells and tissues
I. Title II. Dodds, John H. III. Series
581.87'3 QK725
ISBN 0-412-26890-6
ISBN 0-412-26640-7 Pbk

Library of Congress Cataloging in Publication Data

Bengochea, Teresa.
Plant protoplasts.
(Outline studies in biology)
Bibliography: p.
Includes index.
1. Plant protoplasts. 2. Agricultural biotechnology.
I. Dodds, John H. II. Series: Outline studies in biology (Chapman and Hall)
QK725.B42 1987 581.87'3 86-20714
ISBN 0-412-26890-6
ISBN 0-412-26640-7 (pbk.)

Contents

1

Introduction

Isolated plant protoplasts are 'naked' cells that have had their cell wall removed either by mechanical action or by enzymic digestion. As a result of wall removal the only barrier that exists between the cell protoplasm and the external environment is the plasma membrane. The removal of the cell wall has drastic osmotic consequences for the isolated cell. Rather like animal cells, the isolated protoplasts must be maintained in an osmotically balanced (isotonic) nutrient medium or they will burst.

As will be seen later, the isolation of plant protoplasts is not a particularly new event; as early as the 1890s scientists were isolating protoplasts mechanically and studying effects such as protoplast streaming. What has brought protoplast technology to the forefront of plant biology is the potential application of these structures in non-traditional plant improvement methods.

Extreme care must be exercised by the reader, however, in believing some of the claims that have been put forward for the use of protoplasts in improvement programmes. In the 20 years since the first experiments in routine enzyme isolation of large numbers of protoplasts no real improvement has been obtained in a commercial crop plant. This does not mean that improvement is impossible; it simply indicates that it may be a longer term objective than some people would like to admit. We believe that in the last 3–5 years a new 'realism' has come into this field; and people are now working at experiments that have a chance of success, not at extreme experiments such as fusion of animal and plant cells which were most popular in the 1960s.

Plant protoplasts—as well as being of interest in their own right as potential tools for plant improvement, i.e. protoclones (protoplast-derived variants of clonal material)—are now also being used as components of other improvement programmes; for example, infection with *Agrobacterium tumefaciens* plasmids within genetic engineering experiments.

This volume is intended to give an introduction to the methods used for protoplast isolation and culture, and it indicates the possible uses of protoplasts as tools for plant improvement. It is not intended to be an exhaustive handbook of plant biology, and for this reason each chapter has an exclusive bibliography for those people who wish to follow specific avenues.

The progress of plant biology, especially certain areas of plant tissue culture, has been staggering over the last 25 years. We have moved from a time where tissue culture was perhaps a one-hour lecture of a typical plant biology course to a point where most universities offer a course in tissue culture for their students. Tissue culture has expanded at such a rate that the time has now arrived for individual techniques such as protoplast culture to break off as their own individual areas. Care must, however, be exercised to keep in mind that tissue culture and protoplast culture are not scientific disciplines— they are techniques.

These techniques can, however, be applied to a wide range of biological problems and may offer innovative ways to answer certain questions.

Hopefully, the subsequent chapters will help readers in a wide range of scientific disciplines understand the techniques and allow them to apply those techniques to solve the ever-increasing number of questions posed in plant biology.

2

Isolation and culture

2.1 Introduction

This chapter will describe the various methods used for the isolation and culture of plant protoplasts. It should be noted, however, that an enormous variation exists in the style of experiments and the degree of success of each method. Although certain ground rules can be layed down with regard to these experiments there is still a relatively high empirical approach towards them.

A number of reviews have been made over the last few years covering various aspects of protoplast isolation and culture [1–4]; however, in this chapter emphasis will be placed on the problems associated with each technique.

2.2 Choice of starting material

The process of isolating protoplasts from leaves involves a number of steps; these include surface sterilization, removal of epidermis, enzymic digestion and cleaning of the protoplasts. The physiological state of the leaves is extremely important. The leaves normally chosen for experimental procedures are fully expanded leaves from young plants in a vegetative state (10–14 weeks in tobacco). It is often beneficial to allow the leaves to wilt slightly before removing the lower epidermis. It is inadvisable to subject the mother plant to extreme environmental stresses. The use of systemic pesticides, for example Temik, should be avoided. The lower epidermis that has been removed can be used to isolate epidermal and guard cell protoplasts [5] and the remaining peeled leaf is used for isolation of mesophyll protoplasts.

There are problems associated with the use of intact leaves: it is often inconvenient to remove the lower epidermis, and the leaves have to be surface sterilized. For this reason there has been a move towards using *in vitro* shoot cultures as a starting source for protoplast isolation; these have thin cuticles and do not have to be surface sterilized.

2.2.1 *In vitro shoot cultures*

The use of sterile shoot cultures as a starting material for leaf and stem protoplast isolation has become increasingly popular. One key advantage is that the material is already sterile and does not have to be surface sterilized prior to incubation in the digestive enzyme mixture. Also, within the highly protected and stress-free environment of the culture vessel, the young leaflets are in optimal physiological condition and they develop only a very thin cuticle, thus giving little resistance to the penetration of the enzyme mixture. Sterile shoot cultures have been used as starting material in a wide range of plant species, from potato (Fig. 2.1) and tobacco, and a wide range of tree species. The shoot propagation media varies from species to species, but normally is a Murashige and Skoog salts base, plus a carbon source and agar with the addition of relatively high concentrations of cytokinin and a low concentration of auxin. Sometimes gibberellic acid is added to promote adequate elongation of the shoots [6].

2.2.2 *Callus (cell suspensions)*

Actively growing young cell suspension cultures are excellent starting material for the isolation of large numbers of non-green protoplasts. Normally the cell suspension is filtered prior to treatment with the enzyme mixture to remove any large cell clumps. Old or stationary-phase culture should be avoided as they have a tendency to form dense clumps of cells and giant cells with thickened walls that are difficult for the enzyme mixture to degrade. In some cases the inclusion of a low concentration of cellulase (0.1%) in the cell suspension culture media for 2–3 days prior to protoplast preparation has been advised. This apparently weakens the cell walls and makes the ultimate digestion process simpler [7].

Figure 2.1 *In vitro* potato plantlets, an excellent source of leaf material for protoplast isolation.

2.2.3 Roots and storage organs

Protoplasts have been isolated directly from a wide selection of roots and storage organs. These include potato tubers [8], artichoke tubers [9], onion roots [10] and root nodules [11]. The enzyme treatments are sometimes slower and often a more complex enzyme mixture is required. However, a wide range of root protoplasts have been successfully isolated and cultured.

2.2.4 *Pollen*

The isolation of protoplasts from pollen offers a number of exciting possibilities. The material is readily available in large amounts and is uniform in ploidy. It is ideal material for studies of mutation, cell modification [12, 13], and protoplast fusion technology (*see* Chapter 4). Fusion would cause the formation of a normal diploid instead of doubling the chromosome number (to tetraploid) of the new hybrid. It is also possible to regenerate haploid plantlets from isolated pollen protoplasts [14].

One of the characteristic features of a pollen grain is its highly sculptured and chemically complex cell wall. The wall is covered in a highly resistant coat of sporopollenin, as polymer of carotenoid esters. This coat can be removed by a pretreatment with strong oxidizing solutions. However, even after the removal of the sporopollenin coat the pollen cell wall is a highly complex structure and a complex enzyme digestion mixture is normally used.

If protoplasts are isolated at the pollen mother cell (PMC) or pollen tetrad (PT) stage the wall is less well developed and the protoplasts can be more easily isolated. The enzyme mixture used normally contains high concentrations of B–1.3 glucanase, which is capable of digesting both cellulose and callose. Pollen makes an excellent starting material for protoplasts because it is highly cytoplasmic with a centrally located nucleus. Highly vacuolated material is normally poor starting material for protoplast isolation.

2.2.5 *Other sources*

Protoplasts have with varying degrees of success been isolated from every part of the plant. In addition to the more common sources listed above protoplasts have also been isolated from coleoptiles [15], flower petals [16], aleurone layers [17] and plant cell tumours or galls [18, 19]. The basic isolation technology is the same in all cases.

2.3 Enzymes and osmotica

The structural composition of the cell wall dictates that the digestion mixture should be able to degrade cellulose, hemicellulose and pectin, and in some cases callose. The number of commercial enzyme preparations available is somewhat limited (Table 2.1) and very often they are contaminated with a wide range of biologically active impurities. These impurities may be other enzymes—i.e. proteases, nucleases and lipases—which will have a degrading effect on the plasma membrane of the isolated protoplasts.

Some workers recommend that the enzymes should be purified prior to use on a Sephadex G50 gel filtration column [20], although this practice is not common.

The composition and concentration of the enzyme mixture has a significant effect on the yield of protoplasts from a given tissue.

Table 2.1 Commercial enzyme preparations

Onozuka R10, cellulase
Cellulysin, cellulase
Driselase, cellulase
Macerozyme R10, pectinase
Pectinase
Rhozyme HP150 hemicellulase
PATE (pectic acid acetyl transferase)

The enzyme mixture is normally dissolved in culture media together with an osmotic stabilizer. Calcium (2–10 mM) is a necessary component and phosphate (0.5–2.0 mM) appears to stabilize the isolated protoplasts.

Once the cell wall has been digested away the isolated protoplast is subject to osmotic stress. If an osmotic stabilizing agent is not included in the medium the isolated protoplasts would take in water by the process of osmosis and would eventually burst as there is no cell wall to constrain the cells. Thus, osmotic stabilizers are used to adjust the osmotic potential of the bathing incubation medium, until it is isotonic to the isolated protoplast cytoplasm. The osmotic agents normally used are the sugar alcohols sorbitol and mannitol (13% w/v). Sucrose can be used, but it should be remembered that the isolated protoplasts will use the sucrose as a metabolizable carbon source and the concentration of sucrose in the culture media will fall as the protoplasts are maintained in culture.

Care must be taken not to have too high a concentration of osmoticum as this will cause the protoplast to shrink; when this happens cell wall regeneration is inhibited and subsequent growth is delayed or inhibited.

2.4 Isolation methods

Once one has chosen the appropriate starting material, the next task is the physical isolation of protoplasts.

2.4.1 *Mechanical isolation*

Many people have the idea that the isolation and culture of plants protoplasts is a very new topic. However, the mechanical isolation of protoplasts was first described in 1892 [21]. It is now used only rarely because it is an extremely tedious process that results in the yield of only very small numbers of protoplasts. It does, however, have one advantage over the more popular enzymic methods because there is no possibility of unknown side effects of the enzyme mixture (and their impurities) on the plasma membrane of the protoplast.

The methodology behind the mechanical method of protoplast isolation is shown diagrammatically in Fig. 2.2. The cells are plasmolysed, causing the protoplast to shrink away from the cell wall; a cut is then made across the tissue piece with a scalpel. Some of the cell walls will be cut without damage to the protoplast. When the tissue piece is deplasmolysed the protoplast in the now damaged cell wall will swell and be squeezed out into the bathing culture medium.

2.4.2 *Enzymic isolation*

A typical system for the enzymic isolation of protoplasts is shown in Fig. 2.3. In this particular case the starting material is sterile *in vitro* shoot cultures. In some cases protoplasts are enzymically isolated in a 'two-step' method [22]. First, isolated cells are prepared by treating the leaf or other tissue segment with macerozyme (a pectinase) in 13% mannitol. The isolated cells obtained from this incubation are purified by filtration through a nylon mesh. Secondly, the protoplasts are prepared by incubating these isolated cells in 2% cellulase for about 90 minutes. In the 'one-step' method both enzymes (cellulase + pectinase) are used simultaneously [23]. It is impossible to give fixed rules as to the concentration of enzymes to use and the

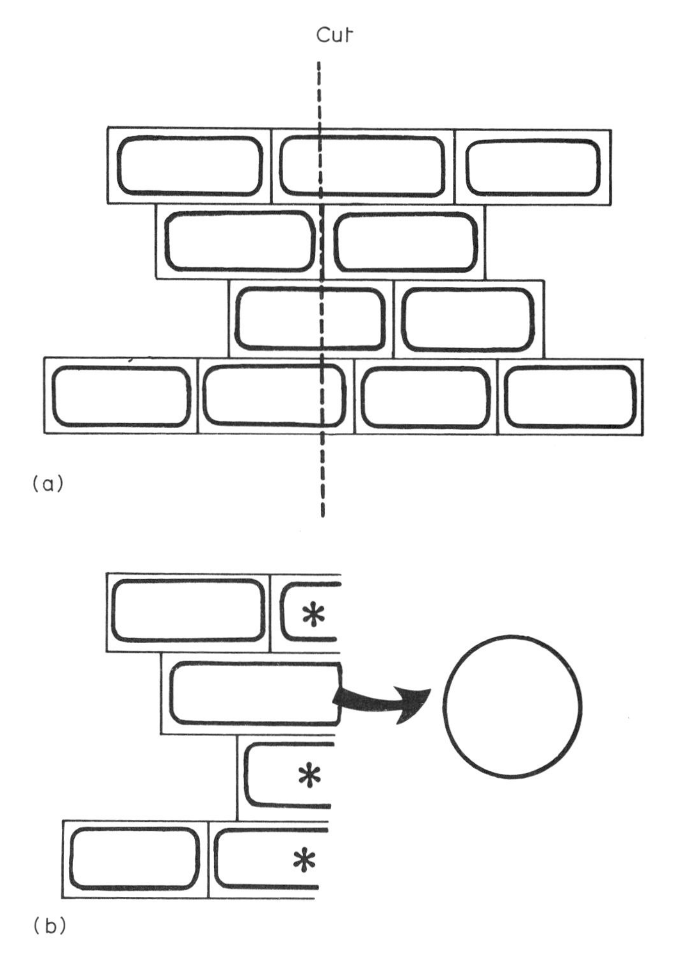

Figure 2.2 Principle of mechanical isolation. When tissue is cut along the dotted line, with say a razor blade, some protoplasts will be released.

times necessary for the incubations as this will vary depending on the type of material under study and its physiological condition. However, using *in vitro* potato shoot as starting material and the 'one-step' method we normally find 2–3 hours at 20–22°C is adequate to release significant numbers of protoplasts.

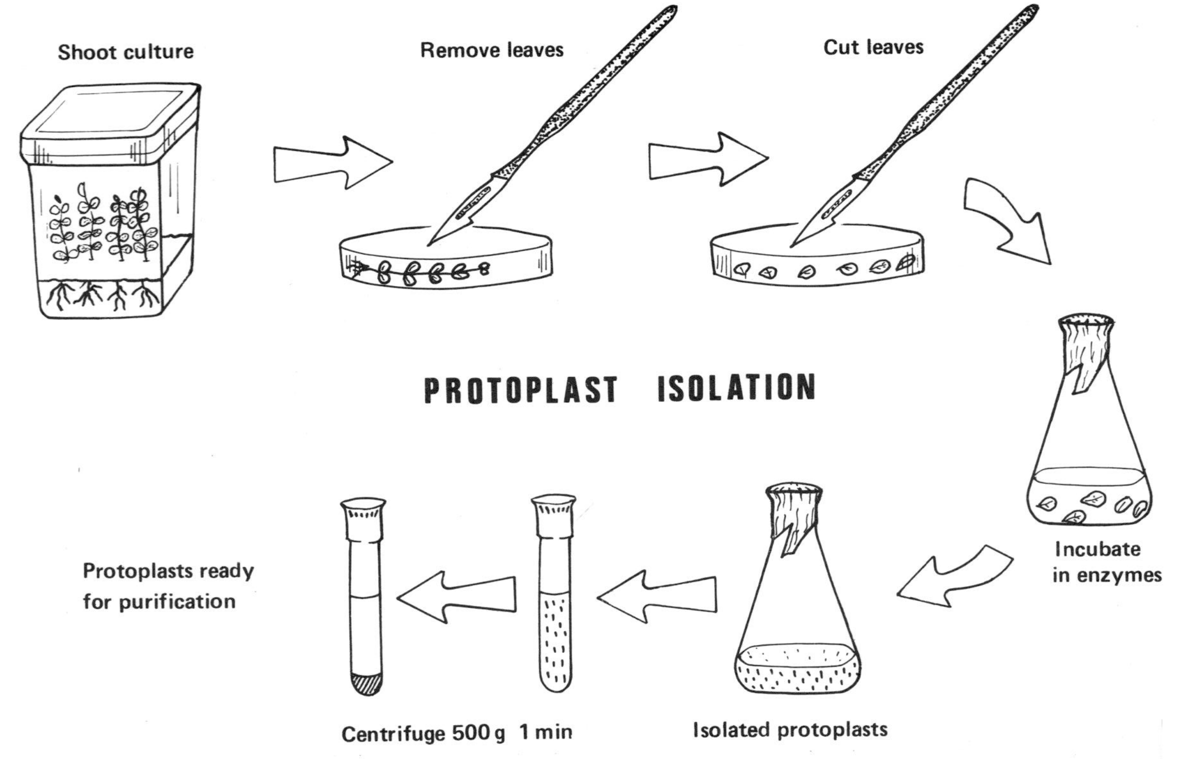

Figure 2.3 Diagram of the principle steps in enzymic isolation of protoplasts. Leaves are removed and cut into fine strips, incubated in enzyme mixture. The enzyme is then removed and the protoplasts are ready for purification.

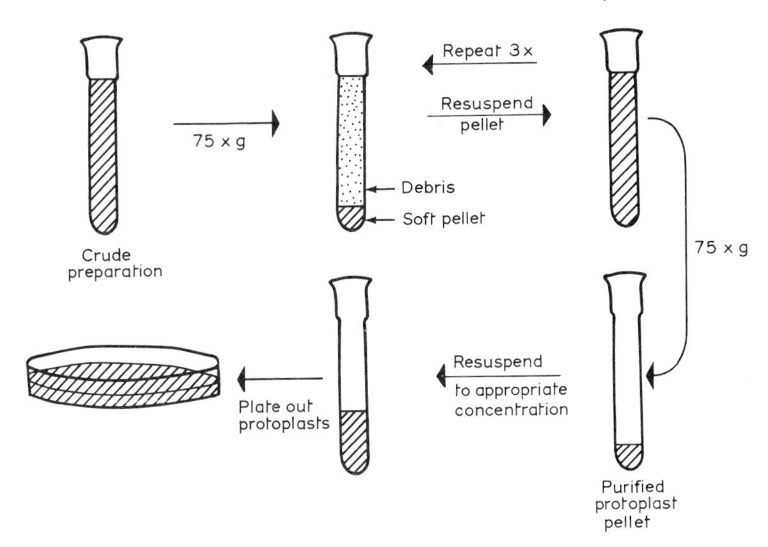

Figure 2.4 Protoplasts can be purified by repeated gentle pelleting and resuspension.

2.5 Purification of isolated protoplasts

The protoplasts isolated in the incubation mixtures described above are present in the media together with a range of cell debris and broken cell organelles. A number of methods have been used to separate the intact protoplasts from this surrounding cellular debris.

2.5.1 Sedimentation and washing

This process is shown diagrammatically in Fig. 2.4. The crude protoplast suspension in normal osmotically adjusted media is decanted into a conical tip centrifuge tube and centrifuged at low speed (50–100 × *g*/5 min). Under these conditions the intact protoplasts form a soft pellet in the tip of the tube. The supernatant, containing the broken cells and debris, can then be carefully pipetted off. The pellet is then gently resuspended in fresh culture media plus mannitol and rewashed. This process is repeated two or three times to eventually give a relatively clean protoplast preparation.

2.5.2 *Flotation*

Because intact protoplasts have a relatively low density when compared with other cell organelles, many types of gradients have been used that allow the protoplasts to float and the cell debris to sediment. A concentrated solution of mannitol, sorbitol or sucrose may be combined with the enzyme/protoplast mixture and then centrifuged at the appropriate centrifuge speed. Protoplasts can be pipetted off from the top of the tube. Some workers prefer to use Babcock bottles (Fig. 2.5) to make the removal of the protoplast fraction easier. The concentration of sucrose used for flotation varies from 0.3 to 0.6 M [24]; however, a detailed analysis of flotation methods for protoplast purification would be useful. The flotation

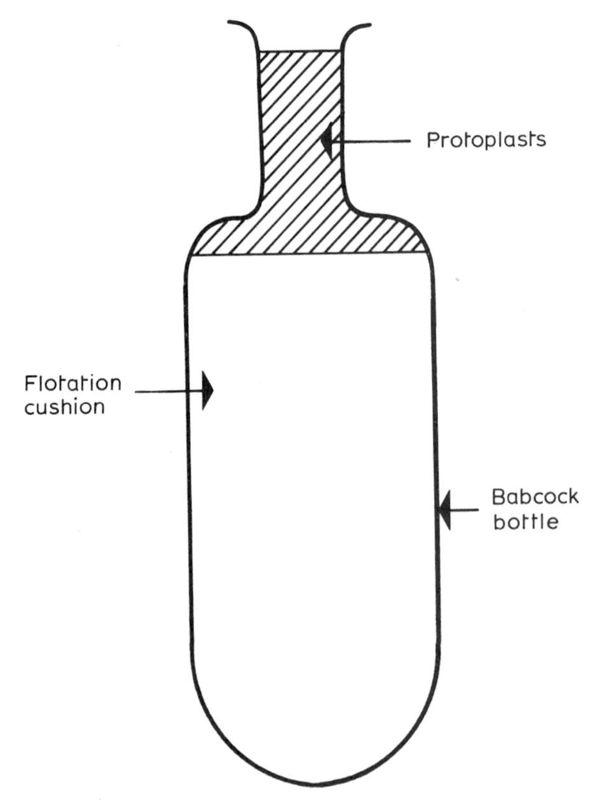

Figure 2.5 Flotation of the protoplasts in a Babcock bottle greatly eases the removal of purified protoplasts from the sucrose cushion.

method causes less loss due to damage than the pelleting and washing method; however, the high concentrations of osmoticum may cause osmotic stress to the protoplasts, delaying wall formation or leading to eventual loss of viability.

2.5.3 Other purification methods

Other more complex pelleting and flotation systems have been used for protoplast purification. Two-step Percoll (polymer of silica) gradients have been used. After centrifugation debris was in the lower half of the gradient and protoplasts floated to the top [25]. A number of discontinuous (layered) gradient systems have been used. Some workers have layered crude protoplast preparation of *Hyoscyamus muticus* onto 20% Percoll. Other workers collected barley protoplasts at the phase border between 0.5 M sucrose and 0.14 M sucrose mixed with 0.36 M sorbitol.

2.6 Protoplast viability and plating density

Having obtained a reasonably pure isolated protoplast preparation, it is important to determine what proportion of the protoplasts are alive or viable and to be able to culture the protoplasts at an appropriate concentration [26].

2.6.1 Protoplast viability testing

In the early studies of this area a number of methods were used: these included observation of cytoplasmic streaming [27], exclusion of Evans Blue dye [28], changes in protoplast size induced by changes in the level of osmoticum [29] and measure of photosynthetic and respiratory activity [30]. Most of these methods have proved to be rather unreliable as techniques for determining viability.

One of the most frequently used methods for estimation of viability is the use of fluorescein diacetate (FDA) [31]. As the FDA accumulates within the plasma membrane viable protoplasts fluoresce green/white. Protoplast preparation treated with FDA (normally 0.01%) must be observed within 5–15 minutes because after this time the FDA dissociates from the membrane.

Another stain commonly used is Phenosafranine (0.01%). This

stain detects dead protoplasts, which turn red in the presence of the dye. Even after 2 hours in stain solution viable protoplasts remain unstained.

Calcofluor white (CFW) can detect the onset of cell wall regeneration. Normally 0.1% v/v solution is used. Wall synthesis is observed by a ring of fluorescence around the plasma membrane [32].

2.6.2 *Plating density*

As is the case with many microbial systems, to achieve sustained growth of a protoplast culture the cells have to be inoculated in the medium at a minimum plating density (mpd). In other words only if above a critical number of cells are inoculated will growth begin. It is believed that cell leakage into the media enriches the media to sustain protoplast growth. In the literature mpd for tobacco protoplants is normally given at between 5×10^3 to 1×10^5 protoplasts/cm^3. Recently, however, there has been great interest in the culture of individual or small numbers of protoplasts. To achieve this goal a number of options are open:

(1) *Small volume cultures.* If the mpd is maintained but the volume is reduced from 1 cm^3 the number of cells in culture also reduces. Using this technique, Gleba [33] cultured small numbers of protoplasts in hanging droplets. In his experiments the droplets were reduced to volumes as little at 2 μl. The same principle is applied in the multidrop array technique described later for the rapid assay of a wide range of different culture media.
(2) *Conditioning of media.* Isolated protoplasts can be cultured at lower densities if grown in conditioned media, i.e. filtered culture media in which plant cells have already been grown.
(3) *Use of feeder layers.* A layer of protoplasts are plated into an agar solidified media. The plate is then subjected to irradiation to inactivate but not kill the layer of protoplasts. This layer is then called a feeder layer. The protoplasts of interest are then overpoured onto the feeder layer at low density. The inactivated feeder layer 'conditions' the layer above and allows the protoplasts of interest to be successfully cultured at between 5 and 50 protoplasts/cm^3.

The normal method of estimating protoplast numbers per cm^3 so that an appropriate dilution can be made with osmotically buffered media is by the use of a Fuchs Rosenthal haemocytometer. Care should be taken, however, that the choice of field depth relates well to the size of the protoplasts. The chambers are normally available with a depth of 0.1 or 0.2 mm. The latter is recommended [34] owing to the fact that the smaller depth might cause the protoplasts to burst, giving an underestimate of number. An alternative counting method, although technically more difficult, is the modification of an electronic Coulter counter. This is an electronic counting device used normally to count yeasts, bacteria and animal cells.

2.7 Methods for protoplast culture

Over the years a wide range of techniques have been developed for the culture of plant protoplasts. The style of the culture often reflects the overall objective of the experiment. If, for example, the objective is to screen very large numbers of protoplasts and pick out only a few colonies that are able to grow then probably a large-scale agar embedded planting method would be used. If, however, the objective is the most appropriate for their experimental needs. These methods wall then perhaps a liquid microdroplet culture would be used.

In the following section a wide range of available culture techniques are described and it is left to the reader to decide which method is the most appropriate for their experimental needs. These methods should not be taken, however, as the ultimate word and the reader must experiment to devise novel methods that may be more applicable to the problem they have under study. The types of culture methods normally used are shown in Fig. 2.6.

2.7.1. Agar embedding

In agar embedded cultures protoplasts may be allowed to regenerate a new cell wall in liquid culture before embedding or may be embedded directly after isolation [35, 36]. The classical method of Nagata and Takebe [37] is to mix the isolated protoplasts with 1.0% agar/culture medium maintained at 40–45°C. Small amounts of the agar (liquid)/protoplasts mixture can then be poured into sterile Petri plates. The isolated protoplasts thus become fixed in a single

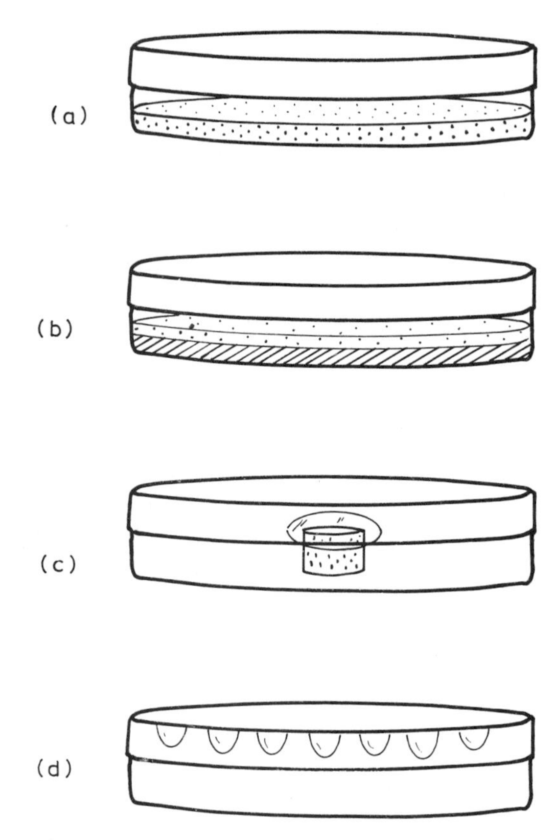

Figure 2.6 Different culture methods. (a) Agar embedding. (b) Liquid culture poured over a feeder layer of medium or medium and inactivated protoplasts of another species. (c) Microchamber constructed of a small plastic ring sealed with oil and a cover slip. (d) Hanging drop cultures.

position in the solidified agar and can be routinely monitored by microscopic analysis.

The above agar embedding technique can be linked to the irradiated feeder layer described above (2.6.2 (3)) and the protoplasts plated at a lower density.

2.7.2 *Microchambers*

The construction of microchambers for the observation of plant cells is quite common [38]. The technique can be modified for the observation of isolated protoplasts. The number of possible designs of chamber are enormous. Fig. 2.6 shows a chamber constructed of a small plastic ring with a coverslip on top. The joints of the system are sealed with mineral oil. The simplest form is to hold a droplet between two coverslips with a ring of mineral oil around to form a seal. The mineral oil prevents evaporation of media; this can be a major problem with protoplast cultures as the osmotic concentration of the media rises and eventually a point is reached where 'seeding' of the mannitol causes crystal formation in the media. Another good sealing substance which is easy to handle is clear nail polish.

2.7.3. *Hanging drop cultures*

Again this technique has been used previously for the observation of isolated cells. The major problem when observing protoplasts is to prevent evaporation of the media. Drops can be placed in the depression of specially prepared hanging drop slides which are then undersealed with a coverslip and oil or the drops are placed on the lid of a Petri dish (Fig. 2.6). The Petri dish contains mannitol solution to maintain the humidity and the dish is sealed with Parafilm.

The size of the hanging drop will be related to the aim of the experiment but normally varies from as little as 2 μl to as much as 200 μl [39].

Further refinements to the hanging drop technique have led to the development of the multidrop array technique described below.

2.7.4 *Multidrop array (MDA) techniques*

This refinement and amplification of the hanging drop technique was devised to allow the rapid screening of a wide range of nutritional and hormonal factors (up to 4900) by a team of people (two technicians) using a small amount of plant material (less than 1 g of plant tissue) [40].

Fig. 2.7 shows diagrammatically how the plates are prepared to give such an array of media types. The system can be described simply as follows:

(1) Prepare a dilution series of two variables A and B, so as to have

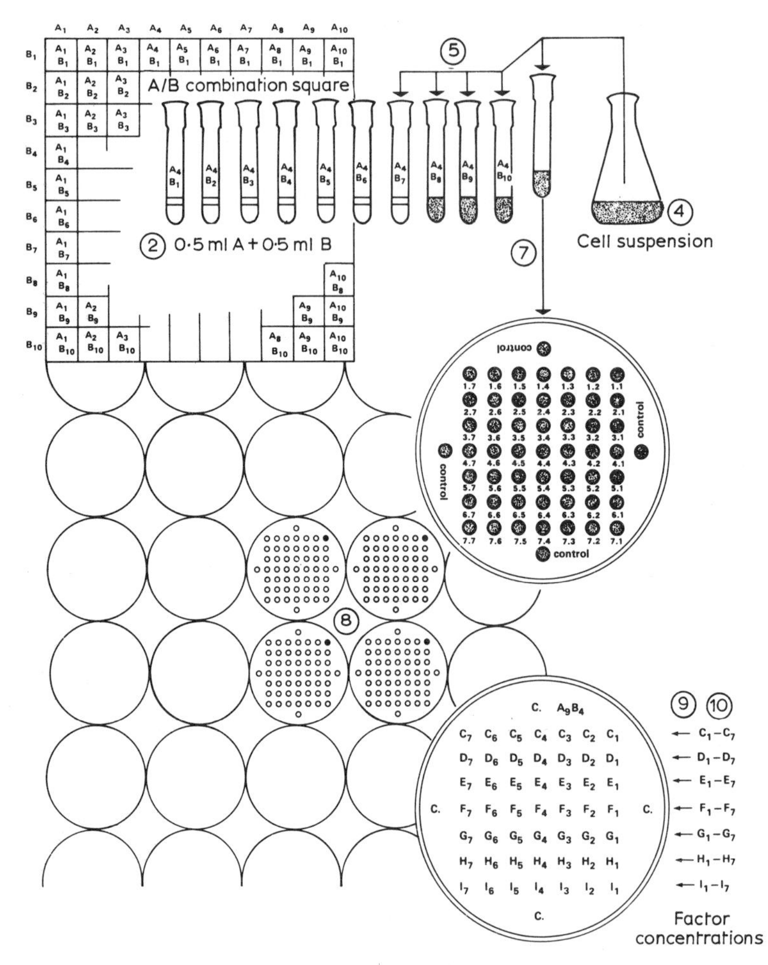

Figure 2.7 Multidrop array technique. (1) Prepare in basal medium a dilution series of the factor(s) to be tested, 5 ml each step. The experiment described uses factors A and B in 10 steps (A_1–A_{10}, B_1–B_{10}), and factors C to I in 7 steps (C_1–C_7 . . . I_1–I_7). (2) Mix 0.5 ml samples of A_1, . . . A_{10} with 0.5 ml samples of B_1, . . . B_{10} to give 100 1 ml samples of all possible AB combinations (A_1B_1, A_1B_2, . . . $A_{10}B_{10}$). (3) Prepare 7-step dilution series of factors C to I in basal medium according to (1), 2 ml each step (C_1–C_7, . . . I_1–I_7). (4) Isolated protoplasts (cells, microspores) and suspend them in basal medium. Adjust their concentration to 20 times the optimal population density (10 ml of this suspension are required). (5) Add 0.1 ml of this suspension to each of the AB factor combinations prepared in (2). (6) Cover air-flow cabinet with wetted filter paper and adjust air flow to low speed. (7)

tubes A1–A10, B1–B10. These can then be mixed to give 100 possible combinations, i.e. A1B1, A2B5, A10B10, etc.

(2) To these 100 different media (1.0 cm^3 each) add 0.1 cm^3 of protoplast suspension prepared in basal media at 20 times the minimum planting density.

(3) For each combination, i.e. A1B1, A1B2, etc. make 49 drops + 4 control drops in order on the lid of a Petri dish and label the dish, i.e. A1B1.

(4) To each drop add 20 μl of another variable media component previously prepared C–C7, I1–I7. Thus each dish contains 49 different media combinations.

(5) A total of 4900 different media compositions each inoculated with cells have thus been prepared on only 100 Petri plates.

(6) 20 cm^3 of mannitol is added to the base of the dish. The lid with attached droplets is inverted into position and the plates are sealed and incubated.

After a suitable incubation period the plates are easily evaluated microscopically (or electronically) to determine which media combination has given the most successful growth. This method has been used to optimize the growth medium in a number of protoplast culture systems [41].

Place 9 cm Petri dishes (uncoated) lid down onto the filter paper. Take off lower part of the dishes. Place 49 + 4 drops (20 μl) of the suspension prepared in (5) in a geometrical 7 × 7 drop arrangement onto Petri dish lid, as illustrated. Use a template for correct arrangement. One dish per AB combination is required (100 plates altogether). When sedimenting protoplasts are used, turn dishes upside down into hanging drop position. (8) Arrange 25 dishes in the cabinet so that corresponding drops are in identical position, as illustrated in step (1) [8]. (9) Add to each drop in position 1.1 20 μ of factor C_1, passing over the series of dishes, continue with C_2 in position 1.2, C_3 in position 1.3 and so on until you complete with factor I_7 in position 7.7. (10) Add 20 μl of basal medium to control drops. (11) Close dishes and turn upside down into hanging drop position. Add 20 ml of mannitol solution to the bottom of each dish. This solution should have approx. 60% the osmolarity of the culture medium used in the hanging drops to avoid drastic changes of the osmotic conditions in the drops. (12) Seal dishes with Parafilm. Repeat instructions (8)–(11) with the next set of 25 dishes. Stack dishes and store them in humidified plastic containers under appropriate culture conditions.

2.8 Cell wall regeneration

The use of protoplasts as a physiological tool for the study of the biochemistry of cell wall synthesis and deposition will be dealt with in a later chapter (*see* Chapter 5). The present section will deal with the time course and developmental importance of wall regeneration.

Protoplasts in culture have been shown to begin new wall synthesis and deposition within a few hours. To complete the synthesis of the wall may take only a couple of days; more commonly it takes a week or more. The presence of the newly synthesized wall can be easily demonstrated by staining it with calcofluor white (0.1%); an alternative would be a plasmolytic method of observation. However, some of the most definitive studies in this area have been the use of a wide range of electron-microscopic techniques such as freeze etch and surface replica preparation. The plant protoplast in culture provides an excellent model system for electron-microscopic study of wall synthesis and deposition. However, some care must be exercised when studying a system of this nature as the *in vitro* system operates under totally different physiological conditions.

The regeneration of the cell wall is the first of a series of critical events that need to take place if the whole totipotential sequence of plant regeneration from a single cell is to be achieved. Without the regeneration of the outer wall it would not be possible for the cell to undergo some of the next steps in the chain of events; that is, synthesis of nucleic acid, replication of chromosomes, mitosis and cell division. It has been shown that in some systems DNA synthesis and mitosis may take place without division, thus forming multinucleate protoplasts; however, these multinucleate cells have never been regenerated to form a whole plant.

2.9 Division and growth

Normally after regeneration of the cell wall the cell undergoes a significant increase in size and the first mitotic division is seen after 5 days (in tobacco). In many cases the cells show a mixture of equal and unequal divisions [42]. This may account for some of the genetic variation found in plantlets regenerated from protoplasts. More will be said about this in future chapters. The second round of divisions is often observed within a week of the first and after 2 more weeks small cell clumps become easily discernible. If these small colonies are left on the same medium as they grow they are subjected to osmotic stress (high mannitol concentration) that restricts their growth or inhibits it altogether. They should therefore be transferred from this medium at a relatively early stage to fresh media without mannitol to sustain further growth into small callus colonies. These callus colonies now provide suitable material for the application of a range of technologies to regenerate whole intact plants. These techniques are described in more detail later (Chapter 3).

2.10 Isolation of subprotoplasts

Subprotoplasts are fragments derived from protoplasts and by definition do not contain the entire content of plant cells. Cytoplasts are subprotoplasts containing most or fractions only of the original cytoplasmic material but are lacking the nucleus. Subprotoplasts containing the nucleus are called karyoplasts or miniprotoplasts. In this case the nucleus is surrounded by a small, but still considerable amount of cytoplasm and the plasma membrane. The term microprotoplast was suggested for subprotoplasts containing not all but a few chromosomes [43]. The term microplast was used to describe protoplast fragments containing only minor fractions of the cytoplasmic material surrounded by an inner membrane of the cell [44]. In contrast to microplasts, subprotoplasts normally are surrounded by the original outer membrane, the plasma membrane. Major emphasis will be given in the following text to enucleated cytoplasts and nucleated miniprotoplasts.

2.10.1 Functional subprotoplasts

Functional subprotoplasts are produced after chemical or physical treatment of protoplasts to facilitate elimination of some of their functions. These morphologically normal looking protoplasts are subprotoplasts only in respect of specific genetic or physiological capacities. Zelcer *et al.* [45] prepared 'functional cytoplasts' by X-ray treatment of freshly isolated protoplasts and thereby inactivated the nuclear genome. Using these treated protoplasts in a fusion experiment they were able to transfer cytoplasmic male sterility independently from the nuclear genome from one tobacco species to another. This technique facilitating the inactivation of one of the fusion partners prior to protoplast fusion has been improved and application of X-ray and iodoacetate inactivation led to a high frequency of cybrid [46]. Laser beam microsurgery may be another method to inactivate specific cellular components [47]. The inactivation of the nuclear genome by physical or chemical means, however, does not exclude absolutely that minor nuclear genomic fragments are 'transferred' following protoplast fusion. In fact, there is evidence for limited intergeneric gene transfer after intensive X-ray treatment of the 'donor' fusion partner [48, 49]. DNA-hybridization experiments will allow more accurate assessment of the degree of limited gene transfer when line specific probes are available.

2.10.2 *Spontaneously formed subprotoplasts*

Protoplasts and subprotoplasts are formed naturally in the pericarps of ripening fruits of some solanaceous species [50–52]. The sap of ripening tomato fruit (6–7 weeks old) contains, in addition to 'normal' nucleated and vacuolated protoplasts, cytoplasts without a nucleus showing different degrees of vacuolation and free vacuoles. All these units obviously are lacking any cell wall. These subprotoplasts are easily isolated by filtration of the juice from the ripening fruit through sieves of different pore size and subsequent sedimentation by centrifugation. The size of the protoplast fragments ranges from 2 to 50 μm in diameter and can be fractionated further on density gradients. Theoretically the very small subprotoplasts could represent fragments containing only mitochondria or plastids, but so far no detailed separation and evaluation has been attempted.

A commonly observed phenomenon during protoplast culture is the so-called 'budding' which gives rise to enucleated subcellular fragments. It appears that during culture the volume of the protoplasts increases and in some cases non-homogeneous cell wall formation results in instability or weakness at the cell surface and leads to the formation of buds. Budding of cultured protoplasts is seen most frequently under conditions that are not favourable for cell division and sustained development. Budding was induced efficiently in *Zea mays* internode protoplasts when cultures were kept for 2–3 days in a medium with relatively high osmotic value of more than 1000 mOsmol/kg H_2O [53]. Multiple formation of cytoplasts from giant protoplasts has been reported by Hoffmann [54]. When cultured protoplasts, isolated from *Nicotiana plumbaginifolia* callus, reached a critical size (about 100 m in diameter) after several weeks in culture without forming a cell wall, the protoplasts started to release cytoplasts into the medium. Under these conditions the separation of the subprotoplasts from the mother protoplast is spontaneous and complete, whereas during protoplast budding there is no complete separation. Budded subprotoplasts are easily separated from nucleated protoplasts by shaking.

So-called microplasts, small subprotoplasts surrounded by an inner membrane of the cell, are released in large numbers when highly vacuolated thin-walled callus cells are ruptured. Watery friable callus with high vacuolated cells can be induced on auxin-containing medium from several plant species. Subprotoplasts obtained from ruptured callus cells or after budding of cultivated cells are of very different size, and enrichment of specific types of subprotoplasts can be accomplished by subsequent fractionating in density gradients [55].

2.10.3 *Plasmolytically induced subprotoplasts*

Plasmolysis of elongated cells frequently causes the shrinking protoplasts to separate into two or more fragments or subprotoplasts. One of these fragments is a nucleated miniprotoplast and the others are enucleated cytoplasts. Subsequent treatment of the plasmolysed tissue with cell wall degrading enzymes produces a mixed population of protoplasts, miniprotoplasts, and cytoplasts (Fig. 2.8).

Due to physical parameters enlongated cells form subprotoplasts

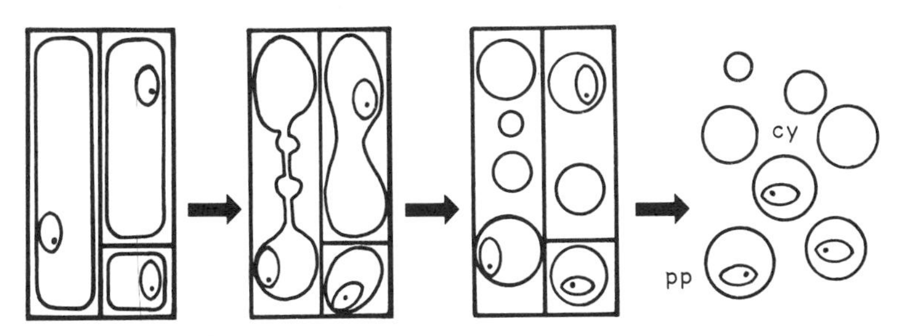

Figure 2.8 When plasmolysed cells are enzyme treated they will often yield a mixture of protoplasts (pp.), miniprotoplasts and cytoplasts (cy).

with higher probability than more isodiametric cells. Bradley has suggested that subprotoplasts would be formed during plasmolysis when the cell length is greater than six times the cell diameter. Observations with onion epidermal cells supported this physical model, where the protoplast is taken as a fluid drop. This simplified model is surely incomplete since subprotoplasts are not formed in all plasmolysed cells with length greater than six times the cell diameter [56].

Vetsya and Bhaskaran [57] isolated subprotoplasts from precultured cotyledonary leaves of *Brassica oleracea* and showed that the yield of subprotoplasts was dependent on the osmolarity of the enzyme mixture used for protoplast isolation. Greater hypertonicity of the enzyme mixture caused more subprotoplast formation. These observations can be explained also in mathematical terms and are related to the increase in surface area to volume ratio in hypertonic medium [58]. Changing the mannitol concentration in the enzyme mixture from 0.44 M to 0.77 M caused an increase in the percentage of enucleated subprotoplasts from *Brassica oleracea* cells from 5 to about 35%. Ten to fifteen per cent of cytoplasts were found in isolates obtained from treatment of *Nicotiana* leaves with cellulase and pectinase. Further enrichment of cytoplasts was achieved by separation of protoplasts and subprotoplasts on a two-step Percoll density gradient, leading to preparations with more than 60% cytoplasts.

2.11 Conclusions

Having obtained satisfactory quantities of isolated intact viable protoplasts that are able to regenerate cell wall, undergo mitotic and cellular divisions and form a callus the next important step is the necessity to regenerate a whole plant. Some of the methods and problems associated with this are described in the following chapter.

References

[1] Cocking, E. C. (1972) *Ann. Rev. Pt. Physiol.*, **23**, 29–50.
[2] Bajaj, Y. P. S. (1977) In *Applied and Fundamental Aspects of Plant Cell, Tissue and Organ Culture* (ed. J. Reinet and Y. P. S. Bajaj), Springer, Berlin, pp. 467–96.

[3] Vasil, I. K. (1980) *Intl. Rev. Cytol.*, **11A,** 246.
[4] Thorpe, T. A. (ed.) (1981) *Plant Tissue Culture. Applications in Agriculture*, Academic Press, New York.
[5] Takebe, I., Otsuki, Y. and Aoki, S. (1968) *Plant Cell Physiol.*, **9,** 115–24.
[6] Wilkins, C. P. and Dodds, J. H. (1983) *Plant Growth Reg.*, **1,** 15–23.
[7] Cocking, E. C. (1960) *Nature (Lond.)*, **187,** 962–63.
[8] Lorenzini, M. (1973) *Compt. Rend. D*, **276,** 1839–42.
[9] Dodds, J. H. (1978) Xylem cell differentiation *in vitro, PhD Thesis.* University of London.
[10] Bawa, S. B. and Torrey, J. G. (1971) *Bot. Gaz.*, **132,** 240–45.
[11] Davey, M. R., Cocking, E. C. and Bush, E. (1973) *Nature (Lond.)*, **244,** 460–61.
[12] Bajaj, Y. P. S. (1975) in *Form Structure and Function of Plants* (ed. B. M. John Commen) Vol.10, Meerut Prakashan, pp.107–15.
[13] Bhojwani, S. S. and Cocking, E. C. (1972) *Nature New Biol.*, **239,** 29–30.
[14] Bajaj, Y. P. S., Davey, M. R. and Grout, B. W. (1975) in *Gamete Competition in Plants and Animals* (ed. D. Mulchany), North Holland, Amsterdam, pp. 7–18.
[15] Hall, M. D. and Cocking, E. C. (1974) *Protoplasma*, **19,** 225–34.
[16] Potykus, I. (1971) *Nature New Biol.*, **231,** 57–58.
[17] Taiz, I. and Jones, R.L. (1971) *Planta*, **101,** 95–100.
[18] Scowcroft, W. R., Davey, M. R. and Power, J. B. (1970) *Planta Sci. Lett.*, **1,** 451–56.
[19] Rollo, F. (1983) in *Genetic Engineering of Eukaryotes* (ed. P. E. Lurquin and Kleinhofs), NATO, Plenum Press, New York, pp. 179–86.
[20] Patnaik, G., Wilson, D. Cocking, E. C. (1981) *Z. Pflanzenphysiol.*, **102,** 199–205.
[21] Klercker, J. A. (1892) *Swenka, Vet. Akad. Forh. Stockholm*, **9,** 463–71.
[22] Otshuki, Y. and Takebe, I. (1969) *Plant Cell Physiol.*, **10,** 917–21.
[23] Power, J. B. and Cocking, E. C. (1970) *J. Expt. Bot.*, **21,** 64–70.
[24] Cocking, E. C. (1972) *Ann. Rev. Pt. Physiol.*, **23,** 20–50.
[25] Kanai, R. and Edwards, G. E. (1973) *Plant Physiol.*, **51,** 1133–37.
[26] Larkin, P. J. (1976) *Planta*, **128,** 213–16.
[27] Raj, B. and Herr, J. M. (1970) *Exp. Cell Res.*, **64,** 479–80.
[28] Kanai, R. and Edwards, G. E. (1973) *Plant Physiol.*, **52,** 484–90.
[29] Evans, P. K. and Cocking, E. C. (1977) in *Plant Tissue and Cell Culture* (ed. H. E. Street), Blackwell, Oxford, pp. 103–35.
[30] Galun, E. (1981) *Ann. Rev. Pt. Physiol.*, **32,** 237–66.
[31] Constabel, F. (1982) In *Plant Tissue Culture Methods* (ed. Lu Wetter and F. Constabel) National Research Council of Canada, Saskatoon, pp.38–48.

[32] Davey, M. R., Frearson, E. M., Withers, L. A. and Power, J. B. (1974) *Plant Sci. Lett.*, **2,** 23–27.
[33] Gleba, Y. Y., Shvydkaya, L. G., Butenko, R. G. and Sytnik, K. M. (1974) *Soviet Plant Physiol.*, **21,** 486–92.
[34] Dodds, J. H. and Roberts, L. W. (1985) *Experiments in Plant Tissue Culture,* 2nd edn, Cambridge University Press, Cambridge.
[35] Bergmann, L. (1960) *J. Gen. Physiol.*, **43,** 841–45.
[36] Thomas, E. (1981) *Plant Sci. Lett.*, **23,** 81–88.
[37] Nagata, T. and Tabeke, I. (1971) *Planta*, **99,** 12–20.
[38] Vasil, V. and Vasil, I. K. (1973) in *Protoplasts at Fusion Cellules somoatiques Vegetales.* Coll. Intl Centre. Nat. Rech. Sci. Paris.
[39] Bawa, S. B. and Torrey, J. G. (1971) *Bot. Gaz.*, **132,** 240–45.
[40] Potrykus, I., Harms, C. T. and Lorz, H. (1979) *Plant Sci. Lett.*, **14,** 231–35.
[41] Potrykus, I., Harms, C.T. and Lorz, H. (eds) (1982) Proceedings of the First International Protoplast Congress, *Experimentia supplement,* **46,** Birkhauser, Basel, Switzerland.
[42] Eriksson, T. and Jonasson, K. (1969) *Planta*, **89,** 85–89.
[43] Bradley, P. M. (1983) *Plant Mol. Biol. Rep.*, **1,** 117–23.
[44] Bilkey, P. C., Davey, M. R., Cocking, E. C. (1982) *Protoplasma*, **110,** 147–51.
[45] Zelcer, A., Aviv, D. and Galun, E. (1978) *Z. Pflanzenphysiol.*, **90,** 397–407.
[46] Menczel, L., Galiba, G., Nagy, F. and Maliga, P. (1982) *Genetics*, **100,** 487–95.
[47] Hahne, G. and Hoffmann, F. (personal communication).
[48] Dudits, D., Fejer, O., Hadlaczky, G., Koncz, C., Lazar, G. B. and Horvarth, G. (1980) *Mol. Gen. Genet.*, **179,** 283–88.
[49] Gupta, P. P., Gupta, M. and Schieder, O. (1982) *Mol. Gen. Genet.*, **188,** 378–83.
[50] Cocking, E. C. and Gregory, D. W. (1963) *J. Exp. Bot.*, **14,** 504–11.
[51] Binding, H. and Kollmann, R. (1976) in *Cell Genetics in Higher Plants,* (ed. D. Dudits and P. Maliga), Akademiai Kiado, Budapest, pp.191–206.
[52] De, D. N. and Swain, D. (1983) in *Plant Cell Culture in Plant Improvement* (ed. S. K. Sen and K. L. Giles), Plenum Press, New York, pp.201–208.
[53] Lorz, H. and Potrykus, I. (1980) in *Advances in Protoplast Research* (ed. L. Freneczy, G. Farkas and G. Lazar), Akedemiai, Kiado, Budapest, pp.377–82.
[54] Hoffmann, F. (1984) (personal communication).
[55] Harms, C.T. and Potrykus, I. (1978) *Theor. App. Gen.,* **53,** 57–63.
[56] Archer, E. K., Landgren, C. and Bonnett, H. T. (1982) *Plant Sci. Lett.*, **25,** 175–85.

[57] Vatsya, B. and Bhaskaran, S. (1981) *Plant Sci Lett.*, **23**, 277–82.
[58] Vatsya, B. and Bhataskaran, S. (1983) in *Plant Cell Culture in Crop Improvement* (ed. S. K. Sen and K. L. Giles), Plenum Press, New York, pp.485–89.

3

Regeneration of plants

3.1 History of plant regeneration studies

The idea that cells could contain all the information necessary for the regeneration of a whole organism is implied in the cell theory proposed by Schwann [1]. At that time, however, the technology did not exist to study regeneration of plants from any single cell other than a newly fertilized zygote, and even this development process takes place within an organized multicellular structure and not as an isolated single cell.

It has long been a standard horticultural practice to regenerate vegetatively propagated plants by taking cuttings. In the early twentieth century a German botanist named Haberlandt took this process as a scientific system for studies of regeneration and attempted to regenerate plants from smaller and smaller cuttings or explants [2]. He was successful at regenerating whole plants from small meristematic pieces but could not obtain any growth of a single cell system. He stated that this inability to regenerate plants from single cells was not because of any fundamental genetic reason, but was simply that an adequate nutrient environment for culture of the material could not be found, and he suggested the use of embryo sac fluids in the culture medium. These statements show an outstanding foresight on the part of Haberlandt and laid the foundations for many later experiments by other workers.

In the mid-1940s a great deal of research concentrated on the use of liquid endosperm in tissue culture media, and in 1941 Van Ovebeek [3] was able to culture isolated *Datura* embryos successfully on a medium enriched with coconut milk. Other scientists rapidly developed the use of liquid endosperm extracts in their media and dramatic effects were found on the growth of both carrot and potato cultures [4, 5].

The regeneration of carrot plantlets from cultured secondary phloem cells of the taproot clearly demonstrated the 'totipotency' of plant cells [6]. At roughly the same time another research group published results showing the formation of somatic embryos of carrot also using a coconut milk based medium [7]. Totipotency (derived from its Latin roots) is the ability to regenerate a whole organism from a single cell. The carrot regeneration experiments cited above finally confirmed the hypothesis proposed by Haberlandt in 1902.

The list of plants from which whole plants have been regenerated from single cells or protoplants has continued to grow and is now well over 100 different species. Some care should be exercised in believing that the concept of totipotency can be applied to all plant cells; in those species where it has been impossible to regenerate whole plants from cultured cells, often the researchers conclude that they simply have not encountered the correct medium. Although in many cases this may be so we should not close our minds to the idea that perhaps some cells may have lost some critical genetic material during differentiation and that those cells simply cannot differentiate into a whole plant. Obviously the latter hypothesis is a far more difficult one to test experimentally than proving totipotency.

Regeneration of plants from isolated cells or protoplants is normally by one of two different pathways, either by organogenesis or embryogenesis. Let us now look at these in more detail.

3.2 Organogenesis

3.2.1 Introduction

Although studies have been made on the *de novo* formation of organs *in vitro* from the early days of plant tissue culture there is still a great lack of information on the induction process. Most investigations centre around the manipulation of three controlling factors: the plant inoculum, the medium, and environmental factors. Organogenesis takes place from a callus, not directly from a single cell. When isolated protoplants are put into culture under appropriate conditions they go through a set series of events. Fig. 3.1 shows these events in regeneration of *Petunia parodii:*

(1) Wall regeneration (*see* section 2.8).

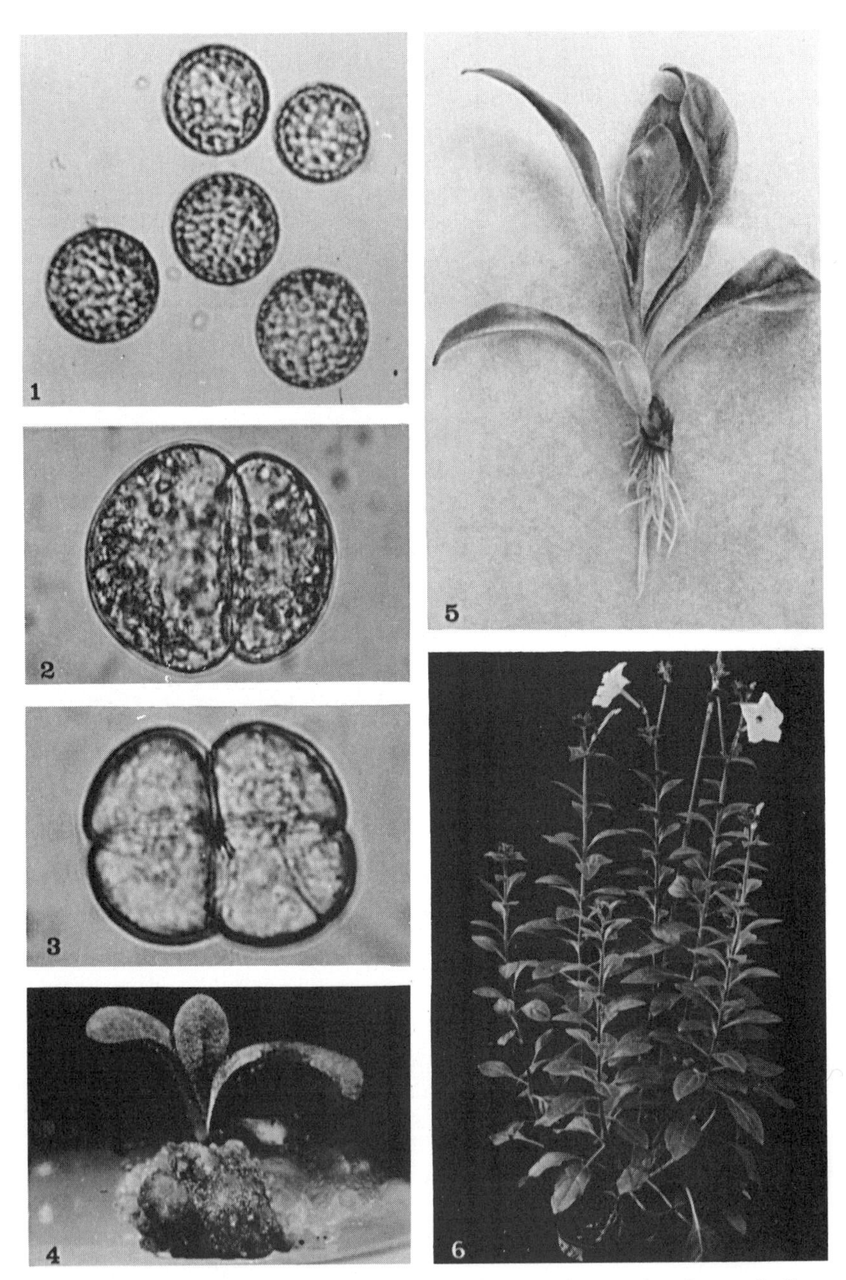

Figure 3.1 Regeneration of *Petunia parodii* plants from isolated mesophyll protoplasts. (1) Isolated protoplasts. (2) Dividing cells. (3) Further mitotic division. (4) Plantlet regeneration from callus. (5) Isolated plantlet. (6) Whole, fertile regenerated plant. (Courtesy J. Power.)

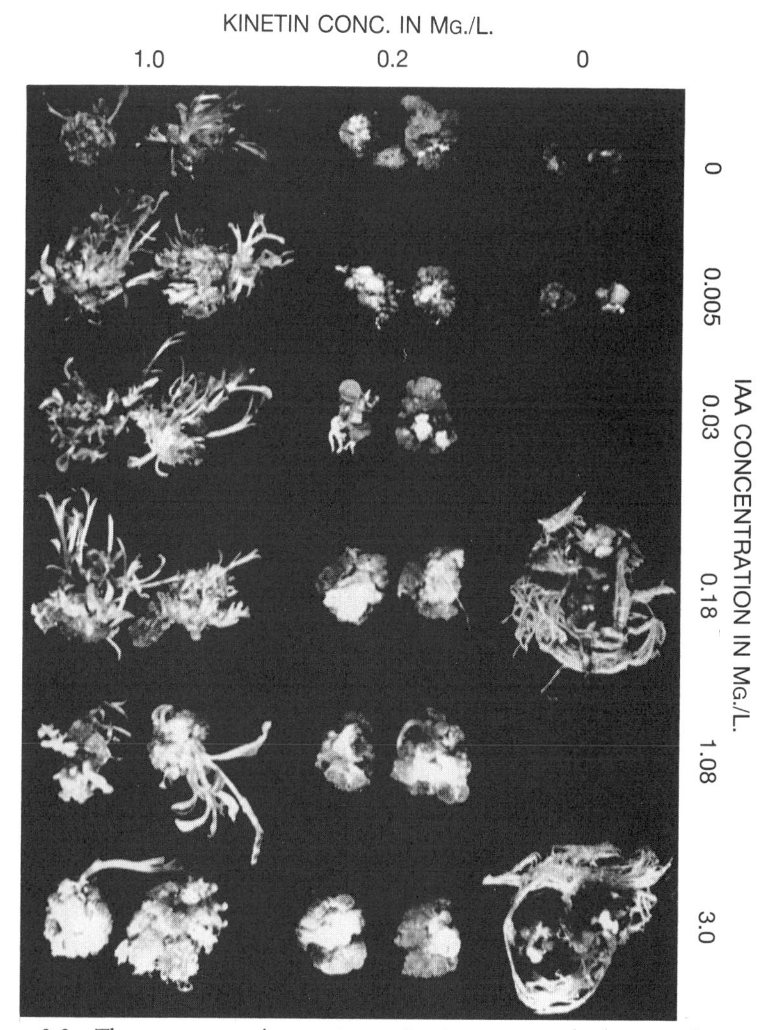

Figure 3.2 The regulation of organ formation in explants of tobacco pith by varying the auxin:cytokinin ratio (courtesy F. C. Skoog).

(2) Early mitotic division and callus formation (*see* section 2.9).
(3) Organogenesis (given appropriate stimulus).

3.2.2. Hormonal control of organogenesis

The first indication that *in vitro* organogenesis could be hormonally regulated was given by Skoog [8]. He found that the addition of auxin to the culture medium stimulated the formation of roots whilst inhibiting shoot formation. The latter effect on shoot production could be partially reversed by increasing both the concentration of sucrose and inorganic phosphate. Later, in a series of new classical experiments, it was found that adenine sulphate (a cytokinin) would promote shoot interaction and that this compound would reverse the inhibitory effects of auxin [9]. Further development of these ideas by Skoog and his collaborators led to the hypothesis that organogenesis is controlled by modifying the balance of auxins and cytokinins within the cultured tissue. A relatively high ratio of auxin:cytokinin would induce root formation whereas a low auxin:cytokinin ratio would favour the formation of shoots (Fig. 3.2).

Another and perhaps more refined system for studying the hormonal control or organogenesis has been the use of cultured epidermal and sub-epidermal explants that are only a few cells thick [10]. The formation of floral buds, vegetative buds and roots has been demonstrated in cultured thin layers of several species by regulating the auxin:cytokinin ratio, the carbohydrate supply and the environmental conditions.

When working on regeneration of a previously studied species it is best simply to check the literature and use the same media and cultural conditions for regeneration. If a new plant species is being studied then it would be necessary to set up a factorial experiment to determine the optimal hormonal and environmental conditions for the induction of organogenesis.

To obtain whole plants that can be transferred to soil the most common regimen is first to apply a low auxin:cytokinin ratio to induce the formation of shoots. These shoots can then be excised and cultured on medium with a high auxin:cytokinin ratio to induce rooting of the cultured shoot. The result is an intact rooted individual plantlet that can then be removed from *in vitro* conditions.

3.2.3 *Cellular origins of regenerated organs*

When a callus fragment is transferred to a suitable medium for organ formation it is clear that not all the cells differentiate into organs. Our understanding of plant developmental biology is still at such a stage that almost nothing is known of the molecular mechanisms that control which cells will differentiate and which will not. It is not even fully clear at this stage whether the regenerated organ is derived from a single cell within the callus or from a small group of cells. In the case of cultured epidermal layers that have a distinct orientation it is known that in most cases the regenerated root or shoot is derived from a single epidermal cell. A knowledge of the cellular history can be important to studies of genetic variation as will be discussed later.

3.3 Embryogenesis

3.3.1 Introduction

The ability of plants to produce embryos is not restricted to the development of the fertilized egg; somatic and haploid embryos can also be induced to form in cultured plant tissues. The phenomenon of somatic embryogenesis was first observed in cultured cell suspensions of carrot, by Steward and his co-workers [11].

Somatic embryos can be induced *in vitro* from three different sources: (1) vegetative cells of mature plants; (2) reproductive tissues other than the zygote; and (3) hypocotyls and cotyledons of embryos. Precisely how these embryos arise have been the subject of numerous investigations. Sharp *et al.* [12] hypothesized that somatic embryogenesis may be initiated in two different ways. In some cultures embryogenesis occurs directly in the absence of any previous callus formation from predetermined embyonic cells. In other cases the development first requires a period of callus formation and the embryos regenerate from induced embryonic cells. Carrot embryos are a good example of this second case.

Embryoids are initiated in callus from small superficial clumps of cells associated with highly vacuolated cells that do not take part in embryogenesis. The embryoid forming cells are densely cytoplasmic with large starch grains (Fig. 3.3). The developing *in vitro* embryoids then pass through a sequence of developmental stages that is exactly the same as seen in the development of a seed embryo. Fig. 3.4 shows the *in vitro* stages of development—that is, globular, heart and torpedo stage then leading to an intact plant.

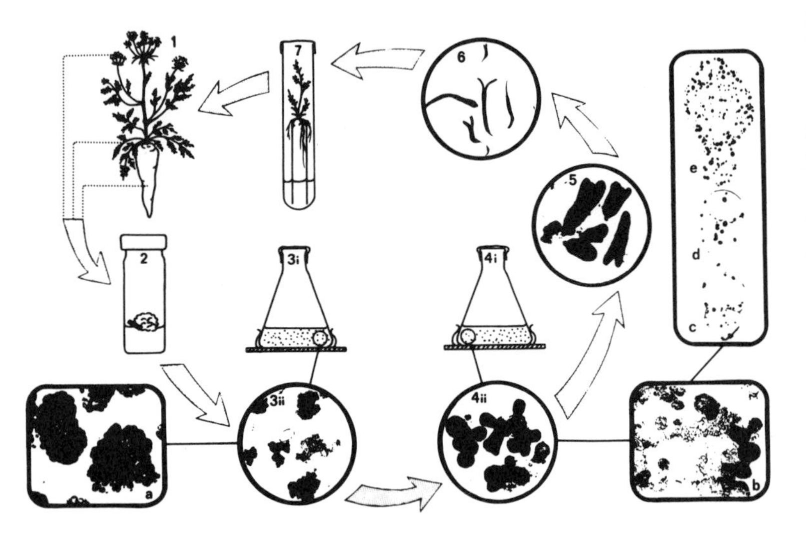

Figure 3.3 Schematic representation of somatic embryogenesis in carrot

(1) Intact plant.
(2) Callus induction.
(3) Cell suspension culture.
(4) Induction of globular embryos.
(5) Heart-shaped embryos.
(6) Mature torpedo embryo.
(7) Regenerating plantlet.

3.3.2 Hormonal control of embryogenesis

The most important chemical factors in the induction medium are auxin and reduced nitrogen [13]. In wild carrot cultures the addition of 10 mM NH_4C1 to an embryogenic medium already containing 40 mM KNO_3 produced near optimal numbers of embryos. Glutamine, glutamic acid, urea and alanine can be used to replace partially the action of NH_4C1 as a supplement. These nitrogen sources are not specific for the induction of embryogenesis, although they are much more effective than inorganic nitrogen compounds.

The role of cytokinins in embryogenesis is somewhat obscure

because of conflicting results. Although zeatin (0.1 mM) stimulates embryogenesis in carrot cell suspensions the process is inhibited by the addition of either kinetin or benzylaminopurine. It has been hypothesized that this inhibition by exogenous cytokinins may result from increases in endogenous cytokinins in the developing embryos.

Supplementing culture media with activated charcoal has proved most beneficial in studies on embryogenesis. However, little is known as to the mechanism of action of the charcoal. It is believed to absorb inhibitory compounds, although it will also absorb to some degree plant growth regulators that are in the medium, thus lowering their effective concentration.

3.3.3 *Synchronization of embryogenesis*

When cell suspensions are cultured in a medium for induction of embryos, the development of the embryos and/or the induction process take place at different rates. The result of this is that often a single inoculum of the culture will contain embryos at all developmental stages, i.e. globular to torpedo. This makes the management of the cultures very difficult. Recent studies [14] have used both chemical and environmental stimuli to try to synchronize the induction and development in the embryos. In carrot this has been quite successful and it is now possible to obtain over 80% of the cultured embryos at the same developmental stage. A further advantage to this synchronization is that it allows more detailed biochemical analysis to be made of the regenerating embryos.

3.4 Handling of regenerated plantlets

Having obtained regenerated plantlets either by the process of embryogenesis of organogenesis the time will often come when it is necessary to transfer these plantlets to non-sterile conditions. This can be a very critical and traumatic period when the plantlet is removed from the carefully controlled environment of a glass test tube to the real world of agriculture. *In vitro* the plantlet had a carefully controlled supply of nutrients, an adequate and controlled water supply, an environment of 100% humidity, and controlled temperature and day length. As the plantlet is moved to non-sterile

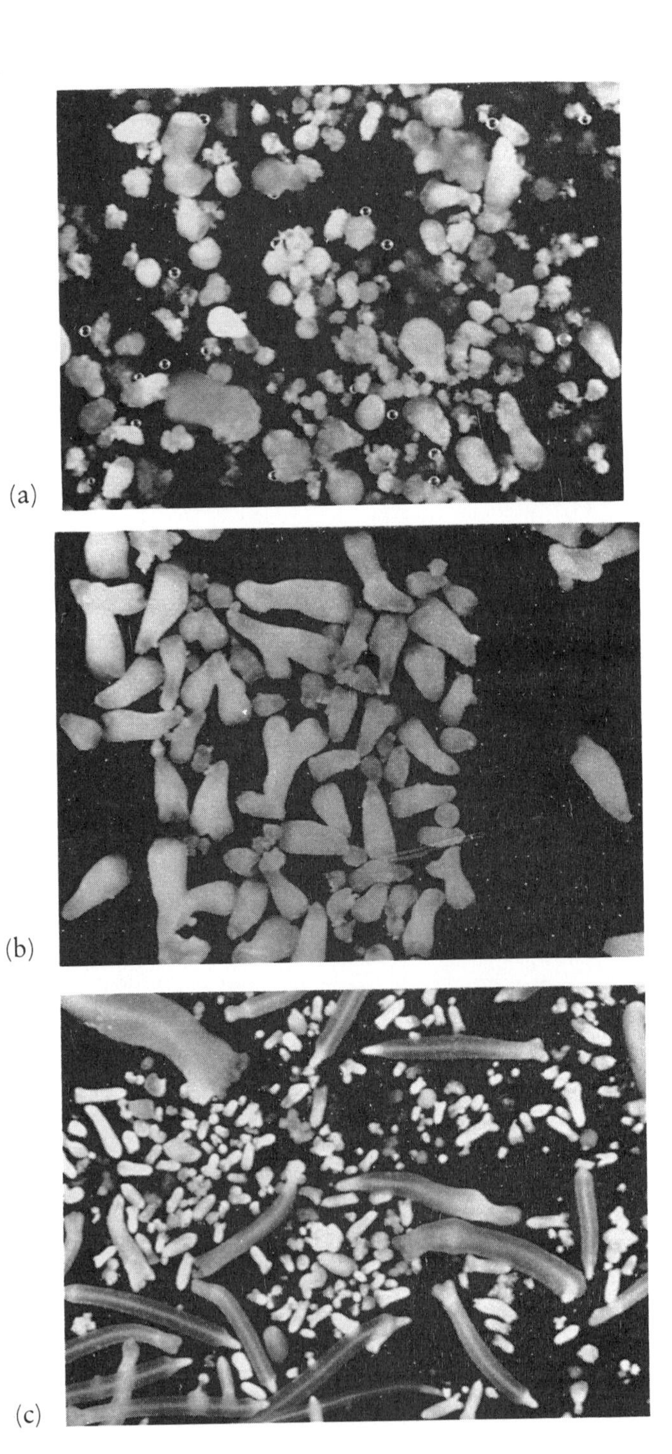

(a)
(b)
(c)

(d)

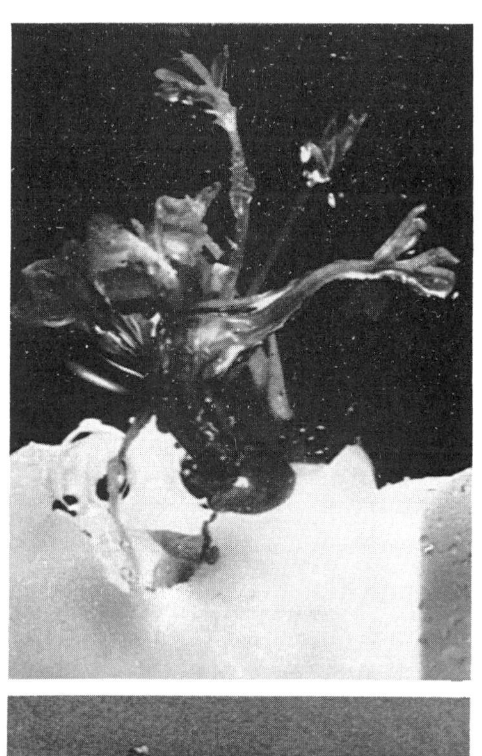

(e)

Figure 3.4 Stages of development of carrot embryoids. (a) Young globular stage. (b) Heat stage. (c) Torpedo stage. (d) Plantlet on filter paper bridge. (e) Mature embryoid derived plantlet (Courtesy L. A. Withers.)

conditions care must be taken to allow gradual release of these careful controls.

3.4.1 *Rooting and water stress*

Plantlets grown in sealed culture vessels are in a constant high humidity environment. Anatomical analysis of the plant has shown that under these conditions often the cuticle covering the leaf surfaces is extremely thin and the root hairs are poorly developed. If these plantlets were suddenly moved to a situation where they are subjected to water stress they would be unable to control their water balance adequately and would die.

It is important, therefore, that *in vitro* plants that are to be transferred to non *in vitro* conditions should be first allowed to develop a good root system and that during the transfer process care is taken to cause the least possible damage to these roots. If the plants have been grown on agar solidified medium this can be removed by gentle washing in a basin of warm water (warm, not hot).

3.4.2 *Transfer to soil*

The plantlets are carefully removed from the test tubes or flasks and any agar medium is carefully removed. The *in vitro* plants are then carefully planted in small pots (Fig. 3.5) and the young roots surrounded with fine sand. It is important that the roots make quick and good contact with the potting mix as the plant will immediately begin to suffer water stress. The planting mix should be well balanced for water retention and should be sterile. Small, sterile peat pots (Jiffy pots) are excellent for this purpose.

If facilities exist the small potted plantlets should then be transferred to a controlled environment chamber, where control of light, temperature and humidity are possible. If this type of facility is not available then the pots can be placed into a home-constructed humidity box. This is a simple wooden frame structure covered with polythene sheet. The plantlets should be shaded from strong sunlight. The polythene box should have a lid that can be removed to 'harden off' the plantlets; for the first couple of days keep the frame closed and then begin to remove the lid for increasing periods of time. During this hardening-off period the plant will develop a normal cuticle layer and a good set of root hairs will develop. Some people water the plants with a 1/5 strength Hoagland's nutrient solution or spray with a dilute foliar fertilizer. The use of this will depend on the quality of the potting mix.

Figure 3.5 *In vitro* potato plantlet being carefully transplanted to small peat pot (courtesy International Potato Center).

3.5 Genetic stability of regenerated plantlets

3.5.1 Introduction to propagation methods

When plants are propagated *in vitro* the plantlets may be derived from the outgrowth of a pre-existing structure such as an apical meristem or an axillary bud or the plant may be formed *de novo* from a cultured cell. For many years tissue culture was heralded as the technique by which it would be possible to produce millions of identical (clonal) individuals within a very short time span by regeneration from single cells. Although it has been proved that large-scale propagation is relatively straightforward in certain species there is still much discussion as to the genetic fidelity of the different propagation systems. One of the most carefully studied systems in this area is the potato, a plant that is normally vegetatively propagated.

3.5.2 Use of the potato for studying genetic variability

The potato (*Solanum tuberosum*) is a plant that shows exceptional plasticity in tissue culture. Plants may be regenerated from almost any organ or explant and from isolated cells and protoplasts.

This plasticity therefore allows the use of this crop as a model for studying variation related to the method of regeneration [15]. Fig. 3.6 shows that a single clone of *in vitro* potato plantlets can be used as a source of material for regeneration of plants using a range of different methods:

(1) Regeneration of plantlets from existing meristems.
(2) Regeneration from stem segments.
(3) Regeneration from leaf discs.
(4) Regeneration from protoplasts.

The regenerated plantlets can then be tested for a number of morphological and biochemical characters to see if they are the same as the original genotype [16].

As a general rule plants that are produced by the outgrowth in culture of an existing differentiated structure, i.e. meristems, are the same morphologically and biochemically as the original genotype.

Plantlets that are regenerated after the formation of a small callus, for example stem segments or leaf discs, will show some features that

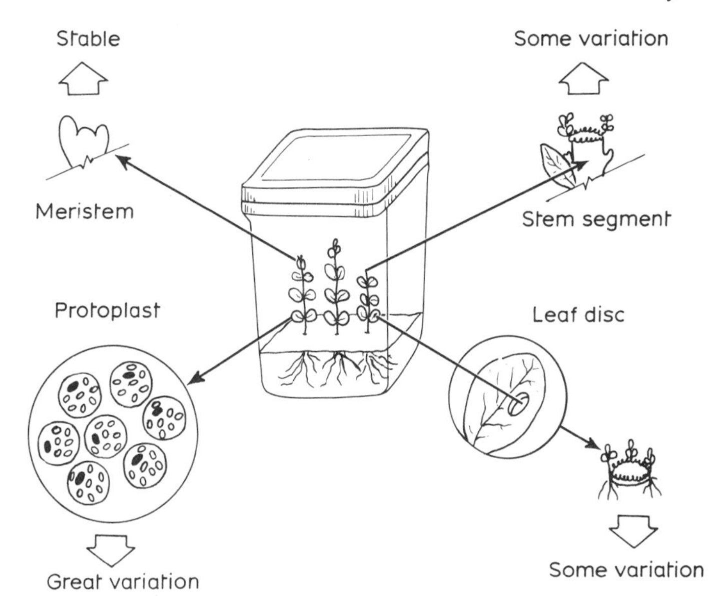

Figure 3.6 Diagrammatic representation of different regeneration methods with potato. Meristem culture, stem explants, leaf discs and protoplasts can be regenerated with different degrees of genetic fidelity.

are different to the original genotype. Fig. 3.7 shows the effect of this type of variation on leaf shape and tuber colour of some regenerated plants in the genotype desirée [17].

If plants are regenerated from protoplasts in potato then a wide range of different plant types are found [18–20]. These protoplast-derived plantlets have been termed 'protoclones' and are currently causing debate among plant breeders.

Whilst the regeneration of plants from isolated protoplast undoubtedly leads to the generation of large amounts of genetic variation (somaclonal variation) [21, 22] there is still heated debate among plant breeders as to the usefulness of this variation. Only time will tell if the use of protoplast culture derived variation is a truly useful tool to plant breeders. However, what is clear is that the use of single cells or protoplasts within a propagation programme is not advisable if one wishes to maintain the characteristics of a certain genotype. For true clonal propagation the only acceptable method that is genetically stable is the outgrowth of existing preformed

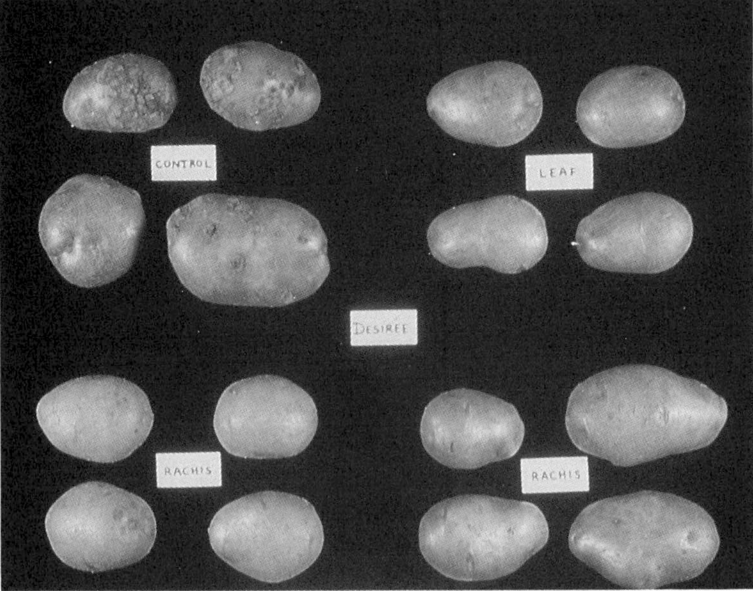

Figure 3.7 Leaf shape and tuber characteristics of callus derived plants from the clone desirée. Note that these are not genetically identical

meristematic structures, that is meristem culture or outgrowth of axillary buds.

To date almost nothing is understood as to the mechanisms of this instability in culture. Many chromosomal abnormalities clearly are caused by culture of protoplasts or single cells. However much more work will be needed to determine the reasons for these instabilities and whether they are random in nature.

References

[1] Schwaan, Th. (1839) *Oswalds Klassiker de Gxakten Wissenschafften*, Englemann, Leipzig.

[2] Haberlandt, G. (1902) *Sber. Akad. Wiss. Wien.*, **111**, 69–92.

[3] Van Overbeek, J., Conklin, M. E. and Blakeslee, A. F. (1941) *Science*, **94**, 350–51.

[4] Steward, F. C. and Caplin, S. M. (1951) *Science*, **113**, 518–20.

[5] Steward, F. C. (1952) *Ann. Bot.*, **16**, 491–504.

[6] Steward, F. C. (1958) *Am. J. Bot.*, **45**, 709–13.

[7] Steward, F. C., Mapes, M. O. and Mearks, K. (1958). *Am. J. Bot.*, **45**, 705–08.

[8] Skoog, F. and Miller, C. O. (1957) *Symp. Soc. Exptl. Biol.*, **11**, 118–30.

[9] Skoog, F. and Tsui, C., (1948) *Am. J. Bot.*, **35**, 782–87.

[10] Tran Trahn Van, K. (1980) *Int. Rev. Cytol.*, **11A**, 175–94.

[11] Steward, F. C. (1952) *Ann. Bot.*, **16**, 491–504.

[12] Sharp, W. R., Sandahl, M. R., Caldas, L.S. and Maraffa, S. B. (1980) *Hort. Rev.*, **2**, 268–310.

[13] Ammirato, P. V. (1983) in *Handbook of Plant Cell Culture* Vol I. (ed. Sharp, Amminato and Yamada) Macmillan, New York, pp.82–123.

[14] Fujimura, T. and Komamine, A. (1979) *Plant Physiol.*, **64**, 162–64.

[15] Dodds, J. H. (1984) *Proceedings of The International Planning Conference on Innovative Methods for Potato Propagation.* (ed. O. T. Page) International Potato Center, Lima, Peru.

[16] Thomas, E. (1981) *Plant Sci. Lett.*, **23**, 81–88.

[17] Jones, M. G. K. (1984) in *Proceedings of The International Planning Conference on Innovative Methods for Potato Propagation.* (ed. O. T. Page), International Potato Center, Lima, Peru.

[18] Shepard, J. F., Bidney, D., and Shahin, E. (1980) *Science*, **208**, 17–20.

[19] Shepard, J. F. (1980) *Plant Sci. Lett.*, **18**, 327–31.

[20] Gunn, R. E. and Shepard, J. F. (1981) *Plant Sci. Lett.*, **22**, 97–101.

[21] Larkin, P. J. and Scowcroft, W. R. (1981) *Theor. App. Gen.*, **60**, 197.

[22] Scowcroft, W. R. and Larking, P. J. (1983) in *Better Crops for Food*, (ed. P. J. Larkin), Pitman, London, p.177.

4

Protoplast fusion

4.1 Introduction

When protoplasts are isolated the only barrier between the cytoplasm and the external environment is the plasma membrane. The lack of the cell wall allows the plasma membrane of two or more protoplasts to come into intimate contact, something which is not possible under normal circumstances. When two protoplast plasma membranes come into contact, under certain conditions they will stick together rather like two soap bubbles. Later, again like two soap bubbles, if given on appropriate stimulus they will fuse together forming a single sphere surrounded by a single membrane [1,2]. Whilst studying mechanically isolated protoplasts, Kuster [3] was able to see occasional spontaneous fusion. However, with the development of enzymic methods for producing large numbers of isolated protoplasts great interest has now developed in the use of protoplast fusion (hybridization) as a possible plant breeding tool. Although positive results in this area leading to agronomically improved plants have been very scarce many people still believe that protoplast fusion offers a useful tool to plant breeders to make crosses between sexually incompatible species for transfer of nuclear or cytoplasmic characters. Protoplant fusion can be used to make crosses within species (intraspecific), between species (interspecific), within genera (intrageneric) and between genera (intergeneric). Let us now look at the wide range of methods available to induce protoplast fusion and to select the products of these fusions.

4.2 Methods to induce fusion

4.2.1 Spontaneous fusion

This as its name indicates is a method that occurs naturally without the inclusion of any fusogenic agent or application of a fusion stimulus. The spontaneous fusion of protoplasts was first described by Kuster [3] in 1902. The major drawback with this method is the rarity with which this event takes place.

Using spontaneous fusion it has been possible to produce a multinucleate soybean protoplast system which showed extensive synchronous nuclear division. Spontaneous fusion has also been described between *Trillium kamtschaticum* and *Lillium longoflorus* [4].

The low frequency of fusion by this method has inhibited its development and more emphasis is placed on the experimental methods of fusion described below (Fig. 4.1).

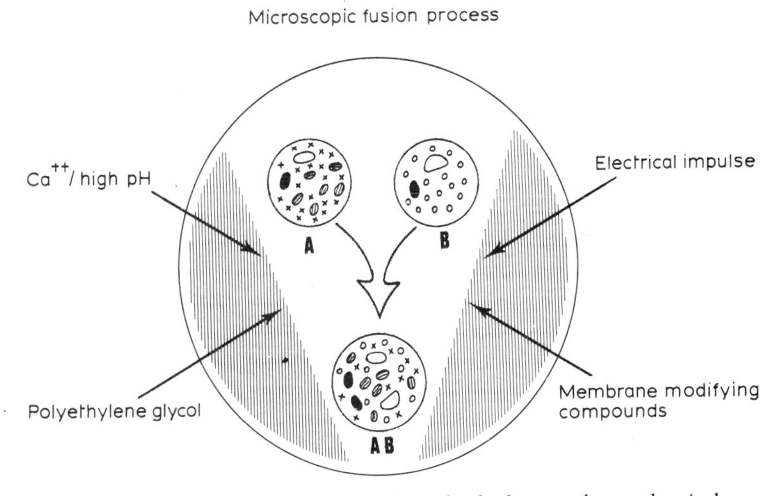

Figure 4.1 The range of experimental methods that can be used to induce protoplast fusion.

4.2.2 Polyethylene glycol

The use of polyethylene glycol (PEG) as a fusogenic agent was first described by Kao at Michaluk [5] and Wallin and co-workers [6].

* 1974

PEG is an organic polymer that can be purchased in different molecular weights. The most effective ranges are PEG 1540 (MW 1300–1600 daltons), PEG 4000 (MW 3000–7000) and PEG 6000 (MW 6000–7500).

To induce fusion the PEG is added to the culture medium in which is suspended a mixture of the protoplasts to be fused. Normally about 25–33% (W/V) of PEG is added. Care must be taken to use the correct amount and molecular size of PEG. Over-addition will cause massive agglutination of the protoplasts whereas addition of too little will cause the protoplasts to adhere but not fuse. The handling of PEG should also be carefully monitored as it is known to be cytotoxic [7].

Normally the protoplast mixture is incubated with the PEG for 10–30 minutes. After this time the PEG is removed by washing, at which time most fusion takes place.

The use of PEG as a fusogenic agent is now very widespread. It is used not only to fuse plant cells but also to fuse together mixtures of animal and plant cells [8].

4.2.3 *Calcium and high pH*

Keller and Melchers [9] described a method of fusion that involves incubating the protoplast mixture in 0.05 M $CaCl_2$ at a pH of 10.5 in 0.4 M mannitol. The protoplasts are centrifuged at low speed ($50 \times g$) for 3 minutes in this mixture. The tubes are then moved and placed in a water bath at 37°C for 45 minutes. Keller and Melchers reported the successful fusion of 25% of the protoplasts. The technique was used for fusion of leaf mesophyll protoplasts and protoplasts from cultured cells. The fusion products could be detected by visual observations (*see* Section 4.3.1).

4.2.4 *Sodium nitrite*

The addition of sodium nitrite ($NaNO_5$) to the mixture of protoplasts has been shown to cause significant fusion. The use of this method was developed by Power and co-workers [10] when they induced inter- and intraspecific fusions using oat and maize protoplasts.

Sodium nitrite induces fusion between highly cytoplasmic, slightly vacuolated and root protoplasts. In general the less differentiated the protoplast the more easy it is to induce fusion.

4.2.5 *Immunological method*

Hartman and co-workers [11] described an immunological method to induce protoplast fusion. They produced an antisera by injecting plant extracts into rabbits. The rabbit would produce a whole range of antibodies against different plant proteins. They then studied the effect of this antisera on fusion of protoplasts from *Bromus. Glycine max* and *Vicia faba*. The antisera caused agglutination followed by fusion of the protoplasts. The fusion products maintained their viability and even some cell divisions were observed.

4.2.6 *Mechanical fusion*

Michel [12] and others [13] have used physical methods such as mechanical pressure to induce protoplast fusion. These methods have attracted little attention as the overall yield of hybrids is still relatively low. However, one advantage of this method is that no chemical agent is used that may have some other effect, for example, mutagenesis, on the cultured cells.

4.2.7 *Electrical fusion*

Zimmermann [14, 15] has recently shown that if protoplasts are placed into a small culture cell containing electrodes and a potential difference is applied, then the protoplasts will line up between the electrodes. If one then applies an extremely short, square wave electrical shock then the protoplasts can be induced to fuse (Figure 4.2). This method has become very popular because of the highly controllable nature of the fusions. In fact it is possible to fuse two single protoplasts. The equipment used for this fusion method has been patented and the equipment can be purchased commercially.

4.2.8 *Other methods*

Kameya [16] described the use of gelatin at 2–5% as a fusogenic agent. He also indicated that potassium dextran sulphate may be a

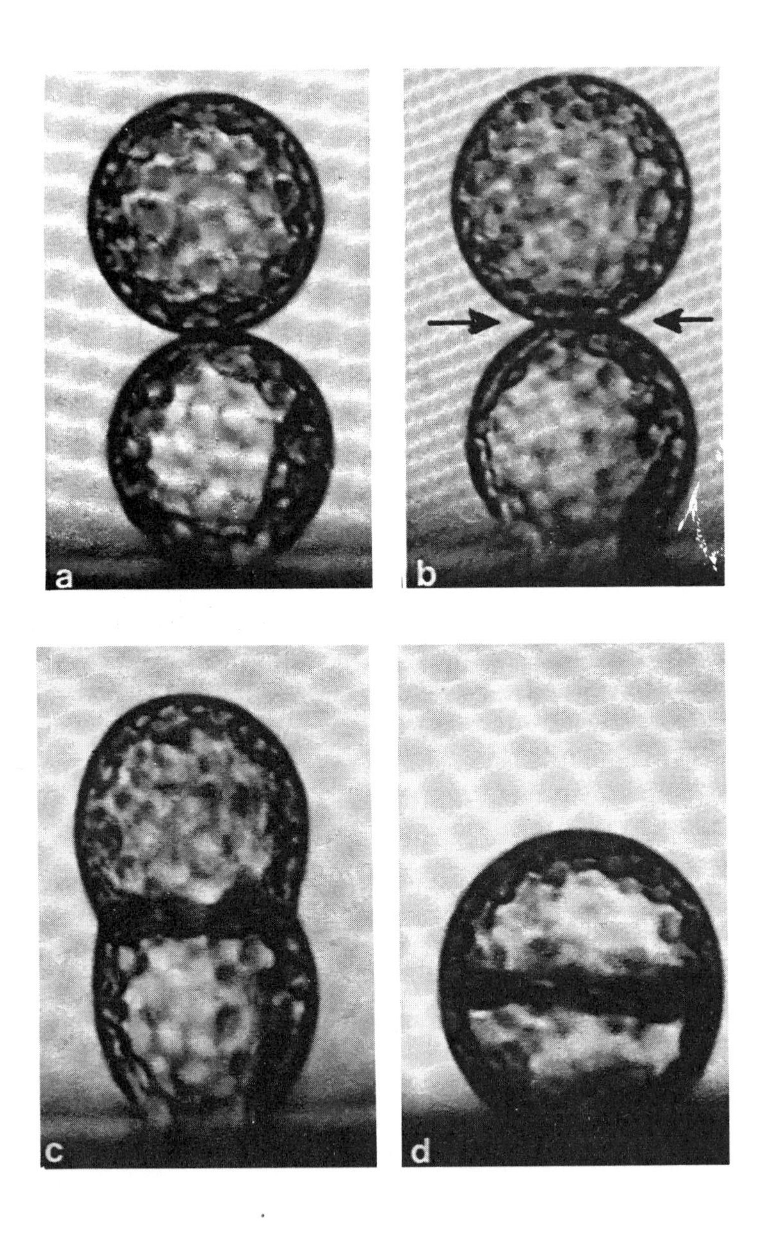

Figure 4.2 Protoplast fusion induced by the application of electrical shock. (Courtesy Dr Zimmermann.)

useful fusogenic agent [17]. Other chemicals that have been used as fusogenic agents with differing degrees of success are concanavalin A, poly-L-ornithine, cytochalasin B, poly-L-lysine, glycerol and dimethyl sulphoxide.

Clearly, a whole spectrum of methods now exists for the induction of protoplast fusion. The great majority of these involve mixing together the two original protoplast types in the presence of the fusion agent and allowing random fusion to take place. As can be seen diagrammatically in Fig. 4.3, the problem with this is that protoplasts can fuse within their own cell type or multiple fusions can take place. Methods have had to be devised that will allow identification of the desired fusion product which is normally the fusion of a single protoplast of type A with a single protoplast of type B. Let us now look at some of the methods used for selection of somatic hybrid selection.

4.3 Selection of fusion hybrids

4.3.1 Visual selection

This method can be used to select fusion products between plants (protoplasts) that have distinct physical characters. For example, Fig. 4.4 shows the fusion between leaf mesophyll protoplasts that contain chloroplasts and a cultured cell that has an anthocyanin-containing vacuole but no chloroplasts. A single fusion product between the two will have both chloroplasts and an anthocyanin vacuole together with two nuclei. The same method can be applied to fusions between leaf mesophyll and cultured cell (achlorophyllous) protoplasts.

The technique of visual detection has sometimes been coupled to microdrop cultures. Fusion products were made between protoplasts of *Arabidopsis thaliane* and *Brassica campestre* [18]. The mixture was then diluted and 500 μl droplets placed into handling drop cultures. The droplets were examined microscopically and those drops containing single fusion products marked so that their future development could be followed.

Obviously, the drawback with this method is that it is very laborious and only a very limited number of fusion products can be selected.

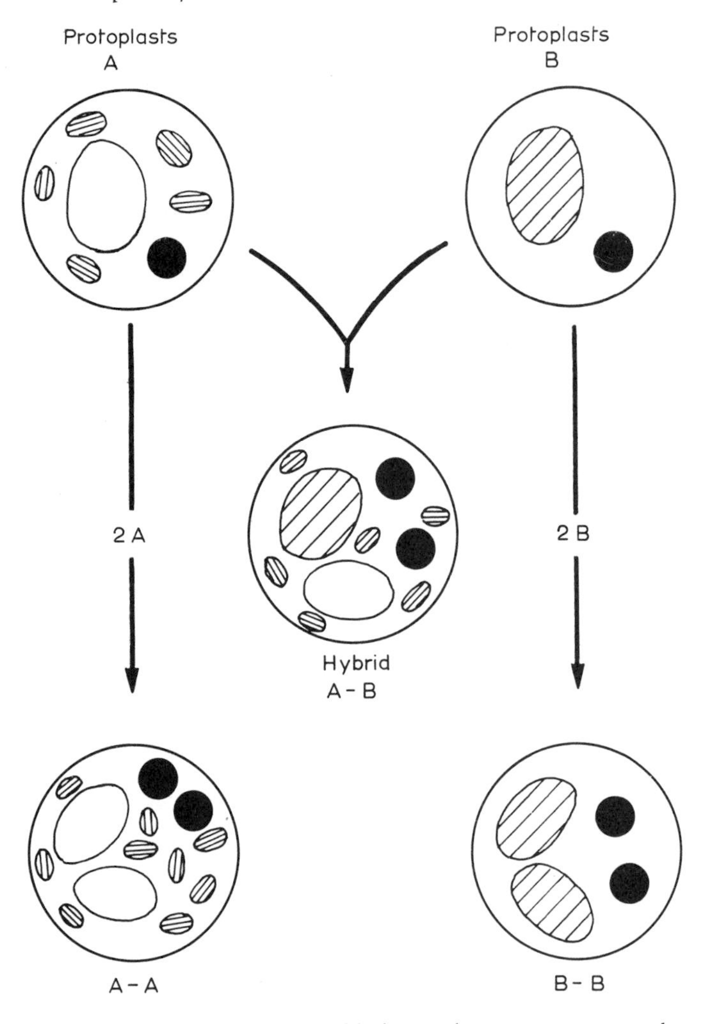

Figure 4.3 Diagram of some possible fusions between two parental types. Normally the desired hybrid would be the single hybrid of the two parents A–B.

4.3.2 Fluorescent labels

The use of fluorescent labelled dyes to study cell fusion was originally described for fusion of animal cells by Keller [19]. The method has been further developed for isolated plant protoplasts [20, 21].

If the two original protoplast cultures are pre-incubated for 12–15

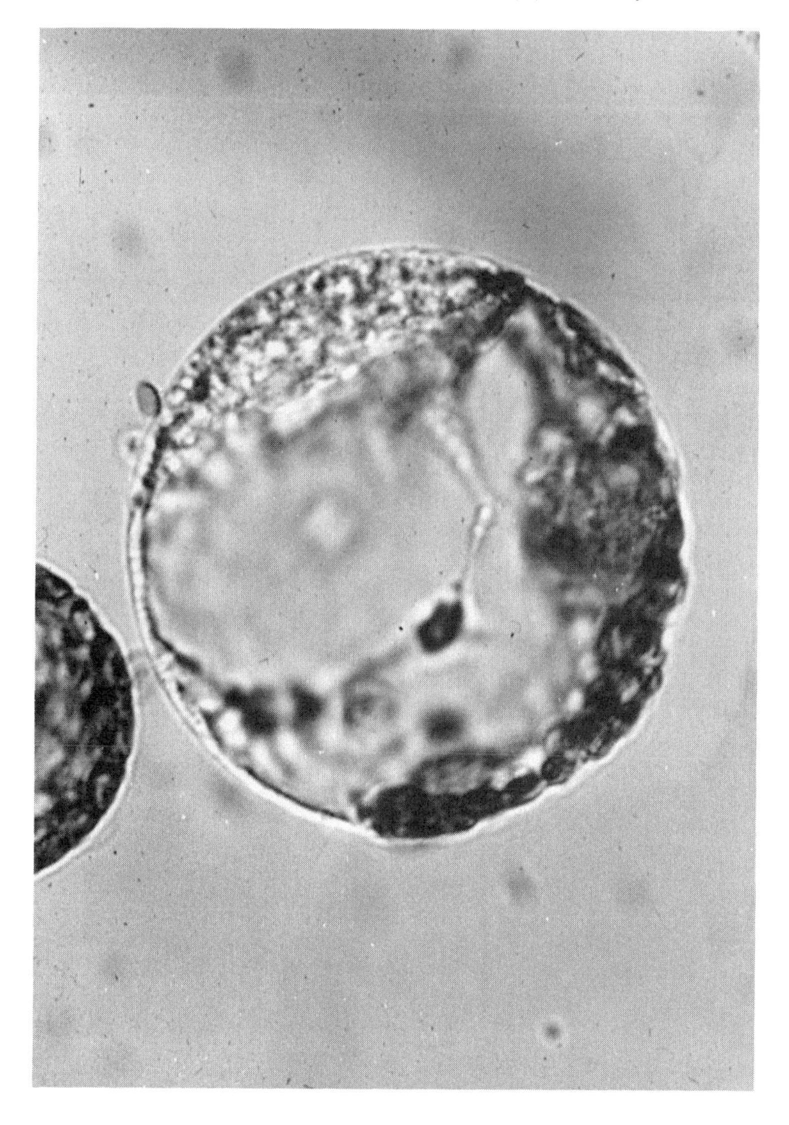

Figure 4.4 Fusion of two protoplasts with distinct morphological characters that allows visual selection of the hybrid. (Courtesy J. B. Power.)

hours, one in octadeconyl aminofluorescein (F18) and the other in octadecyl palamine B (R18) each group of protoplasts takes on a specific fluorescence colour. The dyes are non-toxic and do not affect viability, wall regeneration or growth.

After fusion of the protoplasts fusion products may be identified by their fluorescence characteristics under a fluorescence microscope and a micromanipulator may be used to remove and culture the fusion products. Figure 4.5 shows a mixed culture of fluorescently labelled protoplasts.

The isolation of these labelled fusion products by micromanipulation is still a very time-consuming and skilful operation. A skilled technician would be lucky to isolate 50 hybrids in a day. For this reason attention has turned to the use of mechanical/electronic methods for the separation of labelled protoplasts.

4.3.3 Fluorescence activated cells sorting (FACS)

Protoplasts are fluorescently labelled as indicated in section 4.3.2. After fusion three types of fluorescence can be found: two original fluorescences of the two parental populations and the fluorescence of the fused hybrids. A machine now exists to separate these populations. Fig. 4.6 shows the basic principle of the machine. A droplet containing a single cell is held between two fluorescence detectors. If it is one of the two parent populations the machine charges two

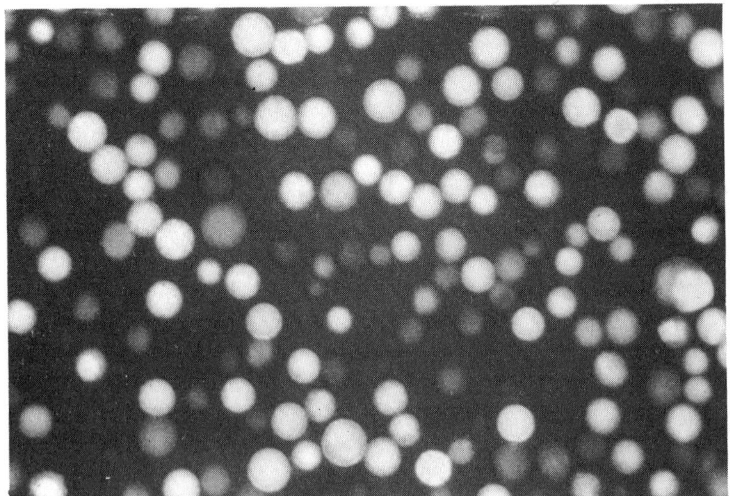

Figure 4.5 Protoplast labelled with fluorescent dyes that allow their easy identification by use of fluorescence-activated cell sorting (FACS). (Courtesy D. Galbraith.)

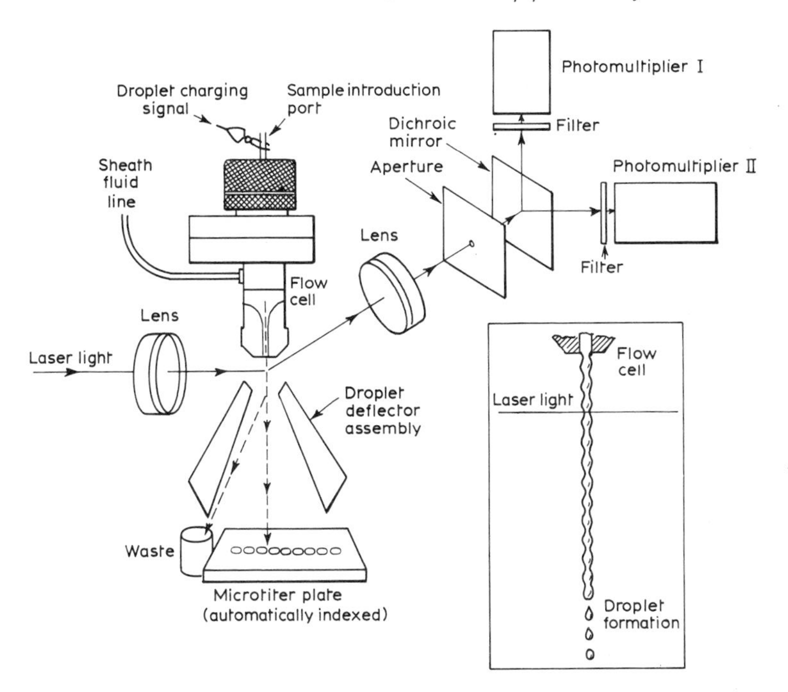

Figure 4.6 Simplified diagram of the principle of a fluorescently activated cell sorter. The droplet containing a single cell is electrically tested for its fluorescence property. By the application of a +ve, −ve or neutral charge to the plates through which the droplet falls it is possible to separate the parental protoplasts and the hybrids.

electrical plates: as the droplet falls it is deflected by the electrical charge. If the fluorescence is of the hybrid no electrical charge is applied to the plates and the droplet fall vertically. Thus the protoplasts are automatically separated into the parent types and the fusion products. The fusion products can then be plated into an appropriate culture system [20]. This method is still very much at an experimental stage. However, it allows the rapid isolation of thousands of fusion products. The major drawback with the technique is its sophistication and cost: a FACS machine costs upward of $100 000.

4.3.4 *Nutritional selection*

In a rather special case, Carlson *et al.* [22] were able to separate

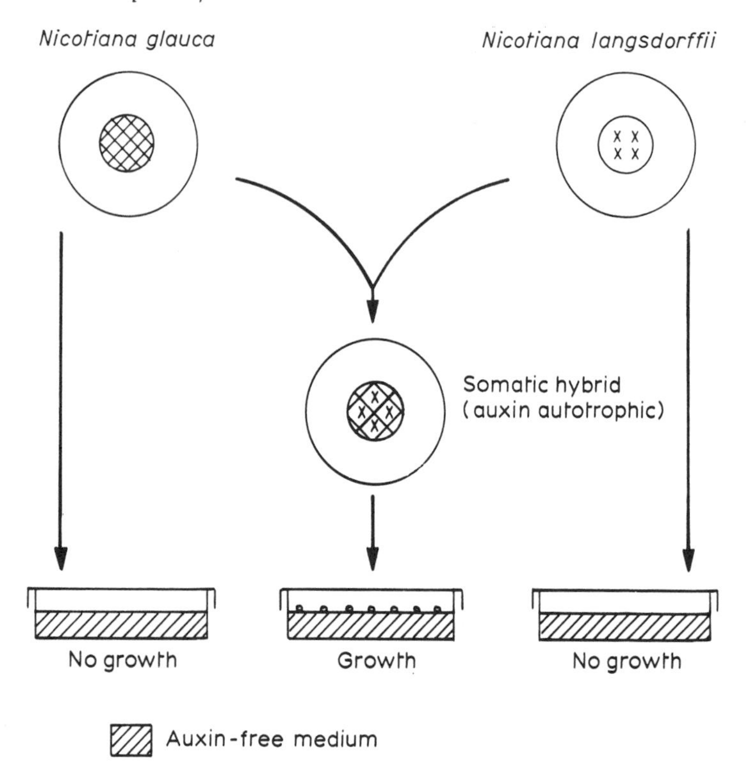

Figure 4.7 Auxin-independent growth of hybrids of *Nicotina glauca* and *Nicotiana langsdorfii* as shown by Carlson. The two parental lines cannot produce auxin and thus do not grown on auxin-free medium. The hybrid cells are able to grow and form callus.

fusion products by their ability to grow on a media that would not permit growth of the parent protoplast lines. It was known that if a normal sexual hybridization was made between *Nicotinia glauca* and *N. langsdorfii* the hybrid plant was tumourous and was able to produce its own supplies of auxin.

As shown in Fig. 4.7 Carlson was thus able to plate out the parent protoplasts and fusion products into a medium that lacked auxin. The two parental cell lines were unable to grow because of the lack of auxin; however, the auxin-producing protoplast hybrid cells were able to grow and form callus. Carlson was able to compare these somatic hybrid plants with those produced by a conventional sexual cross.

4.3.5 Light sensitivity

Melchers and Labib [23] described a very interesting and original method for the selection of hybrids based on genetic complementation. They obtained two chlorophyll-deficient, light-sensitive mutants of *Nicotiana tabacum*. When protoplasts from these two plants were fused the result was a normal green plant. Thus when plantlets were regenerated from the protoplast mixture the two parental clones were light sensitive. However, the hybrid was a normal green plant. This is an elegant method for selection of somatic hybrids. A great need exists for the production of a wide range of selectable mutations, i.e. light-sensitive chlorophyll deficiency to use in these types of studies.

4.3.6 Drug sensitivity and resistance

Power and co-workers [24] used sensitivity to actinomycin D to select fusion products between *Petunia hybrida* and *Petunia parodii*. The two parental species are sensitive to actinomycin D; however, the hybrid is resistant. Power was able to regenerate somatic hybrid plants in the presence of actinomycin D in the medium. Again it would be of great benefit to the field of somatic cell hybridization if a wider range of resistance mutants could be isolated. Some research groups are now attempting to produce large numbers of resistance mutants for use in protoplast fusion studies [25].

4.4 Which plants to hybridize

In the early 1960s plant biologists seemed to cross the line between science and science fiction and proposed some protoplast fusions that bordered on the bizarre. For example, it was proposed to hybridize carrots and cabbage, the logic being that the cabbage would grow and could then be uprooted with a carrot under the soil. The hybridization that achieved the most success as a model system for study was that between the tomato and the potato [26, 27]. Although this has yielded some somatic hybrid plants that have been useful for fundamental studies it has not yielded any plants of agronomic value.

Recently, protoplast fusions have been made between the sexually incompatible potatoes *Solanum tuberosum* and *Solanum brevidens* [28]. The hybrids produced by this somatic fusion are fertile and contain some important resistance genes of *S. brevidens*, i.e. resistance to potato leaf roll virus (PLRV). This method may have use, therefore, not directly to produce agronomically useful plants but as a useful tool for introducing genes into a conventional breeding programme.

4.5 Cybrid formation

When two protoplasts fuse as shown in Fig. 4.8 there are a number of possible outcomes. If the two nuclei fuse then a true hybrid can be formed. However, often the two nuclei will survive independently in the mixed cytoplasms, forming a heterokaryon. Chromosome loss can occur from one or both of the nuclei; if all the nucleus of one parent is lost, what is left is one nucleus in a mixture of both cytoplasms. This product is called a cytoplast or cybrid.

Cybrids can also be prepared by fusing a normal protoplast with

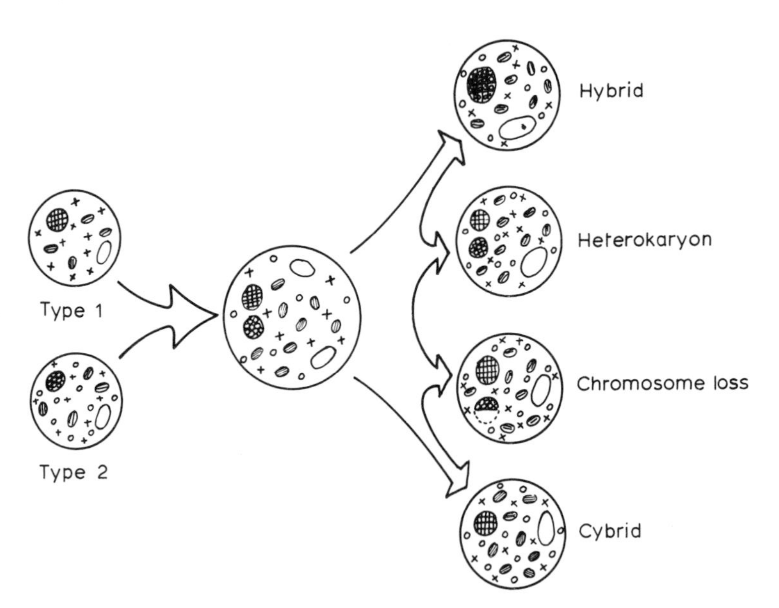

Figure 4.8 Fusion of two protoplants to form a heterokaryon. The nuclei may fuse to form a true hybrid. Chromosome loss or nuclear exclusion will lead to the formation of a cybrid.

an enucleated protoplast. The enucleated protoplast is prepared either by irradiation or by physical removal of the nucleus (see Section 2.10).

In normal sexual production of hybrids the two gametes fuse and produce a zygotic embryo. During this process normally the cytoplasm of the male parent is excluded and there is strict maternal inheritance of cytoplasmic characters [29]. Cybrid production is therefore potentially a useful way of transferring cytoplasmically inherited information.

Presently, researchers are studying the use of cybrids to transfer resistance to some herbicides. This resistance character is carried on the chloroplast genome. Transfer of cytoplasmic male sterility can be controlled by transfer of the mitochondrial genome. The study of cybrids is also important for investigation of possible recombination within these extra-chromosomal genomes.

4.6 Conclusions

To date protoplast fusion has not produced the wonderful research results that it promised in the early 1960s. In fact not one agronomically useful plant has been produced specifically by the application of this technology. However, a great deal of basic scientific information on cell biology and plant incompatibility has been discovered. A number of more specific approaches are now being adopted, such as the transfer of cytoplasmic male sterility. Transfer of male sterility has already been successfully carried out in tobacco [30] and is now being attempted in potatoes [31]. Protoplast fusion may also be a useful tool for the introduction of certain genes from wild species into a conventional breeding programme.

References

[1] Withers, L. A. and Cocking E. C. (1972) *J. Cell Sci.*, **11**, 59–75.
[2] Burgess, L. and Fleming E. N. (1974) *Planta*, **118**, 183–93.
[3] Kuster, E. (1910) *Wilhelm Rouxs Arch Entwick Lungsmech Org.*, **30**, 351–55.
[4] Ito, Y. and Maeda, M. (1973) *Exp. Cell. Res.*, **80**, 453–56.
[5] Kao, K. N. and Michayluk, M.R. (1974) *Planta*, **115**, 355–67.
[6] Wallin, A., Glimelius, K. and Eriksson, T. (1974) *Z. Pflanzenphysiol.* **74**, 64–68.

[7] Mercer, W.E. and Schlegel, R. A. (1979) *Exp. Cell. Res.*, **120**, 417–21.
[8] Davey, M. R., Clothier, R., Balls, M. and Cocking, E.C. (1978) *Protoplasma*, **96**, 157–72.
[9] Keller, W. A. and Melchers, G. Z. (1973) *Natroforsch.*, **28**, 737–41.
[10] Power, J. B., Cummins, S. E. and Cocking, E. C. (1970) *Nature (Lond.)*, **225**, 1016–18.
[11] Hartman, J. X., Kao, K. N., Gamborg, O. L. and Mitler, R. A. (1973) *Planta*, **112**, 43–56.
[12] Michel, W. (1939) *Zellforsch.*, **20**, 230–52.
[13] Scherik, R. V. and Hildebrandt, A. C. (1971) *Proc. Int 1. Center Nat. Rech. Sci.*, **21**, 319–31.
[14] Zimmermann, U. and Schevrich, P. (1981) *Planta*, **151**, 26–32.
[15] Zimmermann, U. and Vienken, J. (1982) *J. Membr. Biol.*, **67**, 165–82.
[16] Kameya, T. (1975) *Jap. J. Gen.*, **50**, 417–20.
[17] Kameya, T. (1979) *Cytologia*, **44**, 449–56.
[18] Gleba, Y. Y. and Hoffman, F. (1978) *Mol. Gen. Fen.*, **164**, 137–43.
[19] Keller, P.M., Person, S. and Snipes, W. (1977) *J. Cell Sci.*, **28**, 167–77.
[20] Galbraith, D. W. and Galbraith, J. E. C. (1979) *Z. Pflanzenphysiol.*, **93**, 148–58.
[21] Patnaik, G., Cocking, E. C., Hamill, J. and Pental, D. (1982) *Plant Sci. Lett.*, **24**, 105–10.
[22] Carlson, P. S., Smith, H. H. and Dearing R. D. (1972) *Proc. Natl. Acad. Sci. (USA)*, **69**, 2292–94.
[23] Melchers, G. and Labib, G. (1974) *Mol. Gen. Gen.* **135** 227–94.
[24] Power, J. B., Frearson, E. M., Hayward, C., George D., Evans, P. K., Berry, S. F. and Cocking, E. C. (1976) *Nature (Lond.)*, **263**, 500–502.
[25] King, P.J. (1982) In *Proceedings of the 5th International Tissue Culture Congress*. Tokyo, Japan (ed. Y. Komamine) Academic Press, New York.
[26] Melchers, G. (1982) In *Proceedings of the 5th International Tissue Culture Congress*. Tokyo, Japan (ed. Y. Komamine) Academic Press, New York.
[27] Melchers, G., Sacristan, M. D. and Holder, A. A. (1978). *Carlsberg Res. Comm.*, **43**, 203–18.
[28] Austin, S., Baer, M. A. and Hegelson, J. P. (1985) *Plant Sci. Lett.*, **39**, 25–81.
[29] Flick, C. E., Bravo, J. E., and Evans, D. A. (1983) *Trends in Biotechnology*, **3**, 2–6.
[30] Aviv, D. and Galun, E. (1980) *Theor. App. Gen.*, **58**, 121–27.
[31] Galun, E. (personal communication).

5

Protoplasts as physiological tools

5.1 Studies of DNA, RNA and protein synthesis

Soon after Nagata and Takebe [1] showed that isolated tobacco leaf mesophyll protoplasts have the capability to regenerate cell walls and divide *in vitro*, Sakai and Takebe [2] reported that these protoplasts incorporated precursors into acid-insoluble fractions of RNA and protein. Several subsequent studies were aimed at investigating further the capacity of isolated protoplasts to synthesize RNA and protein. Thus Wasilewska and Kleczkowski [3] used protoplasts from maize seedling shoot tips to follow RNA synthesis, and reported that gibberellic acid promoted this synthesis and increased the poly $(A)^+$ RNA fraction. As their protoplasts did not show cell division, the incorporation of uridine was terminated after 2–3 hours. Using cotyledon and leaf cucumber protoplasts, Coutts *et al.* [4] reported that uridine was incorporated into ribosomal RNA and that this incorporation was rather low at first but increased gradually up to 1 or 2 days. Such an increase of precursor incorporation was also indicated for tobacco mesophyll protoplasts [5]. Further examination of this and other protoplast systems indicated that the density of the protoplast suspension during culture strongly affects RNA and protein precursor incorporation; that lectins improve leucine, uridine, and thymidine incorporation; and that although the incorporation of these precursors increases steadily during early culture, it is strongly reduced by osmotic stress. An osmotic-stress inhibition of leucine uptake was reported in another system, *Convolvulus* protoplasts, but it should be noted that the latter protoplasts are unable to divide.

In an attempt to compare RNA synthesis in isolated protoplasts with that of intact cells, Kulikowski and Mascarenhas used *Centaurea* cell cultures and protoplasts. They found that protoplasts differed from cells in several respects as to poly $(A)^+$ RNA profiles, levels of rRNA synthesis, and rRNA processing, but protoplasts were similar to cells in the overall rate of RNA synthesis. Unfortunately, in none of the above-mentioned studies was the rate of protein and DNA precursor incorporation accompanied by information on the cell division cycle. Actually, in most cases it is not clear whether the experimental procedure affected the results – which are, therefore, of limited usefulness.

The studies of Galston and his collaborators [6] with oat protoplasts should therefore represent mainly efforts to improve oat protoplast viability by following their capacity to synthesize macromolecules under different culture conditions, rather than to clarify the patterns of protein, RNA, and DNA synthesis during transition from a differentiated cell into a dividing protoplast. Nevertheless, some of the findings with oat mesophyll protoplasts should be noted.

Zelcer and Galun [7] investigated the kinetics of protein, RNA, and DNA precursor incorporation into tobacco mesophyll protoplasts during the first few days of *in vitro* culture. Culture methodologies were modified to give 70% cell division after 70 and 35 hours when incubated at 27°C and 32°C, respectively. Increase in both ^{14}C-leucine and ^{3}H-uridine incorporation began immediately upon culture, or soon afterwards while DNase sensitive/RNase resistant ^{14}C-thymidine incorporation started after a lag of several hours. Moreover, the first peak of thymidine incorporation preceded the rise in percent of divided cells by a few hours.

Ruzicska *et al.* [8] followed the changes in polysomal profiles in protoplasts during culture and compared them with polysomal profiles of comparable intact tissue. They reported that right after isolation there is a drastic reduction of polysomes relative to intact cells, and that with the onset of cell division this profile changes back to normal high-polysomal profiles. Despite the existing RNAses in isolated protoplasts, these could not account solely for all the disappearance of polysomes since exposure of leaves to high osmoticum induces similar RNase levels in their cells without affecting the polysomal profiles.

5.2 Protoplasts for isolation of cell components

The danger of erroneous interpretation of protoplast metabolism as truly representing intact cell processes is hardly encountered when the aim is merely the isolation of cell components. The objective of isolating such components as plasmalemma, nuclei, and chloroplasts from protoplasts, rather than from cells or intact tissues, is in most cases to obtain them by convenient methods which will cause the least destruction. Therefore, if the functionality of these components is of primary importance, the evaluation of the rate and quality of damage (e.g. in nuclei) as well as their unquestionable identification and purity (e.g. of plasmalemma) are important considerations. Most of the problems involved in the isolation of cell components from protoplasts are specific to plant cells because of the latter's rigid cell wall and their specific organelles, e.g. chloroplasts and vacuoles. These problems are very rare in animal cells but are common to all but a few plant cells.

5.2.1 Plasmalemma

Hall and Taylor [9] have recently reviewed the methods for plasmalemma isolation from higher plants. They noted that this isolation is hampered by two major factors: that the high shear force used for cell homogenization will damage the plasmalemma and that good markers are needed to identify the plasmalemma fraction further. The high shear force can be eliminated by the use of protoplasts. Hence protoplasts obtained from either leaf mesophyll or from cell suspensions were isolated by enzyme treatment. Plant protoplasts bind concanavalin A (con A); therefore this lectin can be used as a plasmalemma marker [10]. Taylor and Hall [11] questioned the validity of some plasma membrane markers such as phosphotungstic acid/chromic acid and silicotungstic acid/chromic acid, since these markers may be not specific for plasmalemma, and they further indicated that lanthanum staining, which is rather specific to the plasma membrane, was distributed in all fractions of a sucrose gradient which, according to other markers, was supposed to isolate this membrane. They therefore suggested that further separation improvements are still required before pure plasmalemma fractions can be isolated.

5.2.2 Chloroplasts

The isolation of fully functional chloroplasts from leaves of most angiosperms by mechanical homogenization has usually met with difficulties. Chloroplasts were first isolated from protoplasts by Wagner and Siegelman [12] as a 'by-product' of their method of isolated vacuoles. The protoplasts were ruptured osmotically by suspending them into a phosphate buffer containing Mg^{2+} and dithiothreitol. Chloroplasts were then separated on discontinuous sucrose gradients and were found to be photochemically active. Isolation of cereal chloroplasts was especially problematic; Rathnam and Edwards [13] therefore utilized protoplasts as a chloroplast source. Basically the chloroplasts were obtained by suspending the protoplasts in an appropriately buffered osmoticum (containing bovine serum albumin and dithiothreitol) and passed through a 20 μm nylon net. The chloroplasts were then collected by centrifugation without further fractionation. The protoplasts could be stored in the cold for 20 hr before chloroplast extraction without causing an appreciable loss of chloroplast activity. The isolation of functional chloroplasts from certain plants was found to be increased by specific conditions such as the addition of chelating agents.

A similar method was independently developed by Nishimura for spinach leaf protoplasts, using a syringe needle to rupture the protoplasts and subsequently purifying the chloroplasts by a sucrose gradient. Various enzyme assays (e.g. ribulose-1, 4-biphosphate carboxylase, NAD-triose P-dehydrogenase) were used to verify the purity of the chloroplast preparation. The sucrose gradient step presumably caused some damage to the function of the chloroplasts but resulted in a highly purified preparation. Protoplasts were also used by Nishimura and Beevers [14] to obtain intact plastids from castor bean endosperm. The protocol was basically as reported by Nishimura for spinach leaves, but castor endosperm protoplasts were found to be more fragile than spinach leaf protoplasts.

5.2.3 Mitochondria

High yields of intact mitochondria normally can be obtained from plant tissues and cultured cells by conventional methods; short, high-speed shearing (with sharp knives) is commonly employed with leaf tissue, while disruption in a pressure cell (French press, ca 3000 p.s.i) will give adequate results with calli and cell suspensions. Nevertheless, some of the methods mentioned above for chloroplast isolation from protoplasts may be also useful for mitochondria [15], and the two types of organelles can be separated from the same sample of plant materials. When 0.05–0.1% bovine serum albumin is included and a sucrose gradient is employed for fractionation of the disrupted protoplasts, mitochondria will band quite sharply at a density of 1.18 $g.cm^{-3}$, and they can be identified by their fumarase activity, while chloroplasts will peak at an appreciably higher sucrose density (1.22 $g.cm^{-3}$).

5.2.4 Vacuoles

Vacuoles are very fragile cell components; therefore no large-scale methods of isolation from plant tissue was available before the use of protoplasts. Wagner and Siegelman [16] found that when protoplasts from various tissues are ruptured gently by transfer into a phosphate buffer, they could then be separated by a low-speed centrifugation and be further purified by layering over 5% Ficol containing 0.55 M sorbitol and 1 mM tris-MES buffer. The homogeneity of the vacuole preparation was nicely documented by the use of anthrocyanin-containing tissues, e.g. *Tulipa* petals. Butcher modified the above procedure to analyse the enzymatic content of vacuoles and reported that in mature cells the vacuoles contained some RNases, DNases, and acid phosphatase, but most of the activity of these enzymes, as well as probably all the activity of B-galactosidase, B-glycosidase, or protease, resided in the cytoplasmic fraction, where these hydrolases may be compartmented in small lysosomal-like organelles. Metler and Leonard [17] used tobacco cell suspension protoplasts for the isolation of vacuoles and tonoplast membranes, and discussed previous methods, for the isolation of these constituents. Their procedures differed from previous techniques in two major points: the protoplasts were gently ruptured in a solution containing 0.3 M sorbitol and 2 mM ethylenediamine tetraacetate (EDTA) and the vacuoles were purified on a Ficol step gradient. They concluded that in tobacco suspension cultures the vast majority of acid phosphatase is located in the vacuole. In a further study the tonoplasts were further purfied on a sucrose gradient and were found to have an apparent density of 1.12 g.cm^{-3} and to lack ATPase activity completely. Recently Admon and Jacoby [18] used the cytoplasmic vital fluorescent dye fluorescencein diacetate and demonstrated that cytoplasmic impurities overlay the tonoplasts of vacuoles, because the tonoplast itself was not stained by this dye, the method should be useful in improving vacuole purification.

5.3 Protoplasts for studies on cell wall regeneration

The time course of initial cell wall formation was unclear until recent years. Conflicting reports set the initiation of cellulose fibre formation at various times from 10 min after isolation [19] to 72 hr [20] or even later. When the protoplasts used for studying wall formation were unable to divide in culture, the assumption had to be made that budding indicates preparation for cell division and wall formation. Since cytokinesis seems correlated with wall formation – as will be indicated below – studies on wall formation in these systems should be more meaningful. Moreover, early studies were probably technically deficient; thus the authors had to correct their earlier reported times of cell wall initiation. After these deficiencies were eliminated, Burgess *et al.* [21] could study the time course of wall formation in several species. In tobacco, fibres appeared after 10 hr of culture, as observed by scanning electron microscopy. These fibres could be removed by cellulase and were most probably cellulosic. When the protoplasts were freed from the enzyme and recultured, fibres reappeared without any lag. A few hours after initial fibre formation, a mat of fibres covered the protoplasts' surface, while transfer to unfavourable growth media inhibited fibre formation.

The exact timing of wall formation is of little relevance since culture conditions that facilitate cell division should also facilitate cellulose production. The cellulosic character of the fibres was indicated already by early electronmicroscopic observations on tobacco mesophyll protoplasts. A chemical characterization of cellulose formation in protoplasts from carrot cell suspensions was obtained by Asamizu. These authors found that cellulose was syntheized within 24 hr of culture and that initially short polymers were formed. Cellulose with normal chain length was formed at a later stage but always before the first cell division. Biochemical details on wall formation were subsequently studied in the same system, indicating that during the early period of protoplast cultivation, cell wall components differ from those of intact cells. The arabinose in the non-cellulosic fraction was lower, while mannose, xylose, and glucose were higher in the cultured protoplasts. The latter also released abundant polysaccharides into the medium. Takeuchi and Komamine [22] reported similar observations, indicating differences in wall composition between intact cells and cultured protoplasts derived from suspension cultures. Since no comparable studies were performed with leaf protoplasts, it is not clear whether findings with cultured protoplasts derived from suspension cultures represent the norm, or whether the two systems differ substantially. It should also be kept in mind that caution must be exercised when extrapolating results from wall regeneration in protoplasts to wall synthesis *in vivo*. It may be that although there are similarities the two processes could be under totally independent control mechanisms.

References

[1] Nagata, T. and Takebe, I. (1970) *Planta*, **92**, 301–308.
[2] Sakai, F. and Takebe, I. (1970) *Biochem. Biophys. Acta*, **224**, 531–40.
[3] Wasilewska, L. and Kleczkowski, K. (1974) *Febs Lett.*, **44**, 164–68.
[4] Coutts, R. H. A., Barnett, A. and Wood, K. R. (1975) *Nucleic Acid Res.*, **2**, 1111–12.
[5] Ruesink, A. W. (1978) *Physiol. Plant.*, **44**, 48–56.
[6] Altman, A., Kaiv-Sawhney, R. and Galston, A. W. (1977) *Plant Physiol.*, **60**, 570–74.
[7] Zelcer, A. (1978) *PhD Thesis*. Weisman Institute, Rehovot, Israel.
[8] Ruzicksa, P., Mettrie, R., Dorokhov, Y., Premeez, G., Olah, T. and Farkas, E. L. (1979) *Planta*, **145**, 199–203.
[9] Hall, J. L. and Taylor, A. R. D. (1979) In *Plant Organelles — Methodological Surveys* (ed. E. Reed) Vol 9, Horwood, Manchester, pp. 103–11.
[10] Williamson, F.A., Fowke, L., Constabel, F.C. and Gamborg, O.L. (1976) *Protoplasma*, **89**, 305–16.
[11] Taylor, A. R. D. and Hall, J. L. (1979) *Plant Sci. Lett.*, **14**, 139–44.
[12] Wagner, G. J. and Siegelman, H. W. (1975) *Science*, **190**, 1298–99.
[13] Rathnam, C. K. M. and Edwards, G. E. (1976) *Plant Cell. Physiol.*, **17**, 177–86.
[14] Nishimura, M. and Beevers, H. (1978) *Plant Physiol.*, **62**, 40–43.
[15] Nishimura, M., Graham, D. and Akazawa, T. (1976) *Plant Physiol.*, **58**, 309–14.
[16] Wagner, G. J. and Siegelman, H.W. (1975) *Science*, **190**, 1298–99.

[17] Metler, I. J. and Leonard, R. T. (1979) *Plant Physiol.* **64,** 1114-20.
[18] Admon, A. and Jacoby, B. (1980) *Plant Physiol.*, **65**, 85–87.
[19] Williamson, F. A. Fowke, L. C. Wetter, G., Constabel, F. C. and Gamborg, O. L. (1977) *Protoplasma*, **91**, 213–14.
[20] Burgess, J. and Fleming, E. N. (1974) *J. Cell Sci.*, **14**, 439–49.
[21] Burgess, J., Linsted, P. J. and Bonsall, V. F. (1978) *Planta*, **139**, 85–91.
[22] Takeuchi, Y. and Komamine, A. (1978) *Planta*, **140**, 227–32.

6

Uptake of foreign materials

6.1 Introduction

It was the development of enzymic methods of protoplast isolation in the 1960s that allowed the preparation of large numbers of isolated protoplasts. An isolated protoplast may be defined as a plant cell that has had its outer wall removed; therefore the only boundary between the cell contents and the external environment is the plasma membrane. The removal of the rigid cell wall causes the protoplast to take on a spherical conformation in liquid culture (conformation of minimum energy) and also opens the way to technologies that cannot be applied to other plant cells.

A plant cell surrounded only by a plasma membrane is much more like an animal cell. The cells can be fused together (*see* Chapter 4) to form hybrids in a way that can only be performed with protoplasts. The plasma membrane surrounding the isolated protoplast is also permeable to the uptake of a wide range of other cellular components and even non-biological materials. Let us look in more detail and some experimental uptake systems that have been studied with isolated plant protoplasts (Fig. 6.1).

6.2 Nuclei and subprotoplasts

6.2.1 *Isolation of nuclei and subprotoplasts*

Membrane-bound nuclei may be isolated from a protoplast preparation by treating with cytochalasin B and then centrifuging the protoplasts in a density gradient medium. The process of events is similar to that shown in Fig. 6.2. The relatively high density of the nucleus causes it to be pulled down through the protoplast until it is

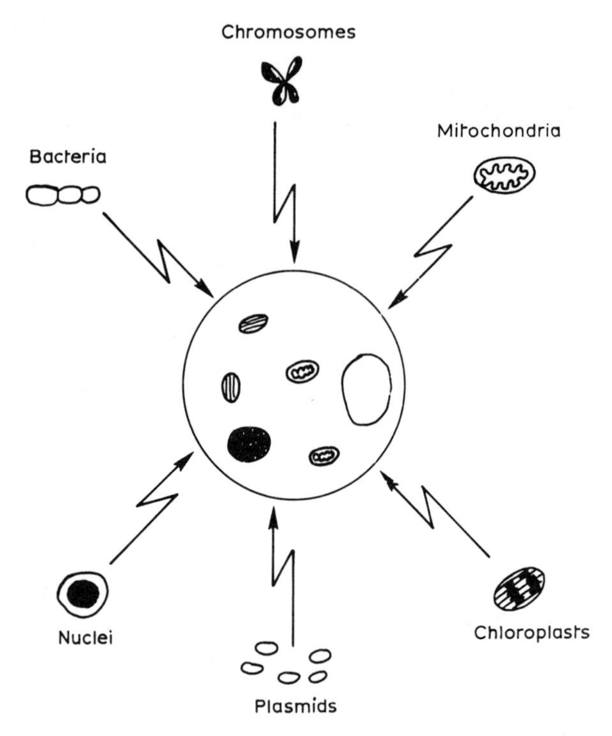

Figure 6.1 A diagrammatic representation of the wide range of biological components that can be taken up by isolated protoplasts.

in contact with the plasma membrane. The membrane is weakened by the cytochalasin B and stretches under the applied stress. The pressure of the nucleus causes extension of the plasma membrane until eventually a fragment buds off. As this happens the two fragments automatically reseal. The result is a nucleus surrounded by a very small amount of cytoplasm (no organelles) and a plasma membrane, and an enucleated protoplast. These enucleated protoplasts can be used as a source of material for cybrid formation (*see* Section 4.6). The isolated nuclei with their membrane covering can then be fused with protoplasts as a method of nuclear transplantation [1].

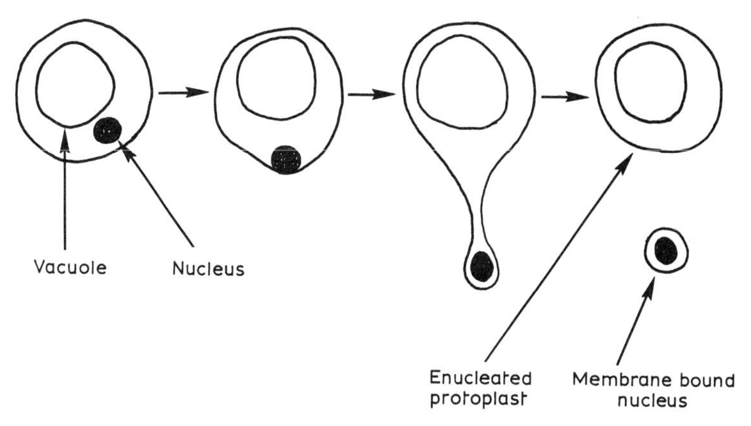

Figure 6.2 Centrifugation method for enucleation of protoplasts. This technique has use not only to produce isolated membrane-bound nuclei but also to produce enucleated protoplasts that are useful for cybrid production.

Although on paper this technique seems a most elegant way of introducing new genetic material it has not yielded any agriculturally useful results to date, and often the fusion introduced isolated nucleus is simply attacked by endonucleases and destroyed.

6.3 Chloroplasts and mitochondria

6.3.1 Isolation of chloroplasts

A commonly used method for the isolation of spinach chloroplasts is modified from Bucke *et al.* [2]. Spinach plants are grown hydroponically in Hoagland's nutrient medium. Plants 2–3 weeks old are routinely used and only the youngest leaves chosen. These are immediately chilled in iced water and exposed for 10 minutes to bright light (250 $W.m^{-2}$). They are then quickly torn into thumb-nail sized pieces and put in a Virtis blender flask containing the isolation solution. Just before the leaf material is put into the blender 2 mM ascorbate is added to the mixture.

As soon as the ascorbate and leaf material are added the whole is blended with the half-thawed medium for 5 seconds and immediately the slurry is filtered through three layers of Miracloth into a centrifuge tube free from all traces of detergent, which might lyse or break the chloroplasts, and centrifuged at 2000 $\times$ *g* for 20 seconds.

The pellet is resuspended in the above medium and its activity measured with an oxygen electrode.

Activities of 50–80 μM oxygen/mg chlorophyll/h are routinely recorded with this method even at pH 6.5, a low pH for photosynthetic activity. Levels as high as 300–350 $\mu M.mg^{-1}h^{-1}$ can be recorded for spinach chloroplasts prepared in the same way but maintained at pH 8.2. This method works well for spinach, pea and *Chenopodium*, but does not yield intact active chloroplasts from all species. Treated with the same degree of care tobacco chloroplasts are normally stripped of their outer membrane and are rapidly inactivated, although the stromal lamellae and grana keep together. There are no methods available that will give intact, active plastids of all plant species and, as yet, those species mentioned above are the only ones from which it can be said that good preparations can be made.

The preparation of chloroplasts from protoplasts has been suggested as a useful method of isolation. Most protoplasts after isolation are actively photosynthetic, though this activity decreases over a period of hours. Preparation of chloroplasts from protoplasts is suggested as being by simple lysis of the protoplast in hypotonic solution and subsequent release of the chloroplasts. Often such a method leads to the formation of groups of plastids, often accompanied by mitochondria, embedded in small pieces of cytoplasm, the whole being surrounded by a membrane, presumably of either tonoplast or plasmalemma in origin. When one speaks of uptake in this case one is really referring to fusion events.

6.3.2 *The induction of chloroplast uptake into protoplasts*

Several methods for the induction of chloroplast uptake into plant protoplasts have been published, all of them depending either on some form of osmotic effect drastically plasmolysing the protoplast or on hypothesized changes in the charge pattern on the plasmalemma. Carlson [3] found it unnecessary to use any agent to promote chloroplast uptake but this is not the general finding.

Potrykus [4] described two methods for the transplantation of chloroplasts of Petunia. In the first method protoplasts and chloroplasts were centrifuged in alternating layers together with 0.03% lysozyme, and then subjected to a hypotonic shock. The second consisted in incubating the isolated single cells of Petunia, prepared by pectinase treatment, with 2% cellulase in 0.35 M mannitol, pH 5.4 at 20°C for 60 min. When the first protoplasts appeared, the single cells were transferred into a suspension of chloroplasts in 2% cellulase in 0.21 M sodium nitrate (equiosimolal with 0.35 M mannitol). The mixed suspension was incubated at 20°C for 15–30 min in a swing-out rotor at low speed at about 10 × *g*. The pellet was resuspended every 5 min. Once all the cells had formed protoplasts the mixed suspension was washed clean of cellulase and free chloroplasts and the protoplasts examined. The possibility that the so-called uptake under these circumstances may actually be fusion as a result of the preparation of chloroplasts from the breakage of protoplasts in the presence of sodium nitrate has already been mentioned. The second of these two methods gave the best uptake but the condition of the chloroplasts after 15–30 min in 2% cellulase at 20°C is questionable. Isolated chloroplasts are normally maintained at, or near, 0°C to retain their activity, and an incubation of this length at 20°C is not calculated to maintain their integrity of metabolic activity, especially since cellulase preparations frequently contain appreciable amounts of protease.

Bonnett and Eriksson [5] used a method involving plasmolysis by polyethylene glycol (PEG) to induce uptake of *Vaucheria* chloroplasts by carrot protoplasts. A dense mixture of protoplasts and chloroplasts was made up to a volume of 0.5 ml with protoplast culture medium. A volume of 1.5 ml of the solution containing 3 ml of protoplast culture medium and 7 ml of a 56% aqueous solution of PEG was added to the protoplast–chloroplast mixture. The final concentration of PEG was about 30%, a concentration known to promote cell aggregation and cell fusion without reducing cell viability. After 10 min the mixture was diluted to 10 ml with a solution of 0.1 M $CaCl_2$ in 0.3 M sorbitol, and centrifuged for 3 min at 150 × *g*. The pellet was resuspended in 10 ml of the same solution, recentrifuged and resuspended in 2 ml of the protoplast culture medium.

6.4 Bacteria

Attempts to obtain uptake of intact bacterial cells into protoplasts stemmed from earlier observations on the uptake of particulate materials into isolated higher plant protoplasts. The plasmalemma of an isolated protoplast should not be thought of as a tightly stretched surface, but as a membrane capable of altering its shape by developing outgrowths as well as invaginations. Extensive ultrastructural studies have demonstrated the frequent occurrence of vesicles within the cytoplasm of isolated protoplasts. Each vesicle is surrounded by a single membrane which in many cases originates as an invagination of the plasmalemma into the cytoplasm. The open end of the invagination then seals as a result of membrane fusion to release the vesicle into the cytoplasm. Vesicles vary considerably in size and are thought to be of two types, namely plasmolytic and endocytotic. Plasmolytic vesicles are often several μm across and originate as gross infoldings of the plasmalemma into the cytoplasm as the protoplast contracts during plasmolysis. Endocytotic vesicles are smaller and arise both during wall removal and following protoplast isolation. The extent to which endocytosis normally occurs in intact plant cells is uncertain, and direct evidence for this phenomenon in plants has come mainly from studies with isolated protoplasts. Removal of the cell wall appears to stimulate activity of the surface membrane, the extent of endocytosis depending on the protoplast system and the constituents of the plasmolyticum used to stabilize the naked cells. As endocytosis generally involves an initial adsorption onto the plasmalemma of the material to be taken up, then the surface charge of this material in relation to the charge of the plasmalemma is critical. Polycations such as poly-L-ornithine can alter the charge on the surface membrane of the protoplasts and may be used at low concentrations (1–2 μg. ml^{-1} to stimulate uptake.

Ultrathin sections of isolated protoplasts suggested that particles considerably larger than 0.25 μm could be accommodated within some of the cytoplasmic vesicles of isolated protoplasts. Indeed, some vesicles were considered large enough to contain whole micro-organisms. This hypothesis was investigated by studying the uptake of bacterial cells (*Rhizobium leguminosarum*; Rothamsted Culture Collection Cat. No. 1007) into protoplasts of pea (*Pisum sativum* cv. Little Marvel) leaf mesophyll cells. The reasons for this choice of experimental material were two-fold. First, it was considered that since this bacterium normally induces nodules effective in nitrogen fixation on roots of the host plant, this micro-organism would be the most compatible bacterium to use with cells from other parts of the same legume species. Second, in the event of successful uptake, the behaviour of bacteria in the cytoplasms of the protoplasts could be studied during subsequent culture of the naked cells.

Experiments were performed to determine whether *Rhizobium* could be taken up into isolated pea leaf protoplasts by endocytosis. The *Rhizobium* cells were grown to shake cultures in liquid mannitol yeast-water medium, and exponentially growing cells harvested for use 3 days after inoculation. After washing to remove the culture medium, the bacterial cells were suspended at a density of approximately 2.0×10^8. cm^{-3} in a plasmolyticum (25% w/v sucrose solution) containing freshly isolated mesophyll protoplasts (5.0×10^5/ml). The protoplasts were prepared from fully expanded leaflets of 5–7-week-old pea plants using a mixture of 5% w/v meicelase with 5% w/v macerozyme in 25% w/v sucrose (pH 5.8) for 20 h at 25°C in the dark [6]. The protoplast were incubated with *Rhizobium* for periods of up to 12 h, but uptake of bacteria did not occur even when the naked cells were treated with polycatonic stimulators of endocytosis. This experiment confirmed that endocytotic uptake into protoplasts is restricted to small particles, as suggested by the earlier experiments of other workers.

The method which resulted in the successful uptake of *Rhizobium* into pea leaf mesophyll protoplasts involved entry of the bacteria into plasmolytic vesicles [7]. In the method described, leaflets excised from young pea plants were wetted by immersion in 1% v/v sterile 'Teepol' for 3 min, and surface sterilized in 1.5% v/v sodium hypochlorite solution containing 10–14% available chlorine (10 min). The sterilant was removed by eight successive washes with sterile water. The lower epidermis was removed by peeling with fine forceps, and the leaf pieces were incubated with their exposed mesophyll in contact with a filter-sterilized enzyme mixture of cellulase and pectinase. 1.5 g fresh weight of peeled tissue was floated on the surface of 5 ml of enzyme contained in a 5 cm Petri dish. The enzyme mixture and incubation conditions were the same as those previously described for the isolation of pea mesophyll protoplasts except that the enzyme contained *R. leguminosarum* at a density of approximately $2.0 \times 10^8.ml^{-1}$. The bacteria were harvested from exponentially growing cultures immediately before use. Protoplasts released during the enzyme incubation were freed of leaf debris by straining through a fine wire gauze, and collected by centrifuging the protoplast–enzyme–bacterium suspension (225 × g; 5 min). The film of intact, floating protoplasts was washed to remove enzymes and excess bacteria from the surface of the protoplasts by re-suspending six times in fresh plasmolyticum followed by centrifugation (225 × g; 5 min). Protoplasts were fixed, embedded, and sectioned for electron microscopy by a routine procedure.

Ultrastructural observations showed the presence of *Rhizobium* localized within membrane-bounded vesicles in the cytoplasm of some of the isolated pea mesophyll protoplasts. Many of the vesicles containing bacteria were deep within the cytoplasm, frequently internal to the other organelles such as chloroplasts and mitochondria, and also within transvacuolar cytoplasmic strands. This location of the membrane-bounded vesicles suggested they were completely closed to the exterior of the protoplasts, with little possibility of them being merely invaginations into the cytoplasm. Similar results have also been obtained using another legume mesophyll protoplasts system. Mesophyll protoplasts were isolated from expanded simple leaves of 8-day-old cowpea (*Vina sinesis* cv. Blackeye) seedlings in the presence of *R. japonicum*, under incubation conditions similar to those described for the isolation of pea mesophyll protoplasts. Bacterial cells were again found localized within membrane-bounded vesicles in the cytoplasm of the isolated cowpea mesophyll protoplasts.

6.5 Blue–green algae

Experiments have been performed [8] where isolated protoplasts are incubated with the blue–green nitrogen-fixing alga *Gloecapsa*. Also the nitrogen-fixing alga *Anacystris* can be taken up. The method normally used is to co-incubate the algal preparation with the isolated protoplast preparation in the presence of 25% polyethylene glycol, when high Ca concentration is added the protoplasts begin to engulf the algal cells. The method of uptake has been shown to be by envagination of the plasma membrane. It is noteworthy that even by this method only about 1% of the protoplasts take up algal cells and the future life of these cells is very limited. We believe that to insert nitrogen-fixation genes to plants will probably not be a very satisfactory method.

6.6 Viruses

Cocking [9] observed the uptake of tobacco mosaic virus by tomato fruit protoplasts and studied the initial stages of infection by the electron microscope. Since then it has been reported that protoplasts from other plants can be infected by TMV [10], cowpea chlorotic mottle virus, potato virus X and cucumber mosaic virus [11]. These studies have demonstrated that TMV/RNA is taken up by pinocytosis through the plasmalemma, and that the infection is stimulated by poly-L-ornithine. The virus enters the cytoplasm of the protoplast and multiplies. The intracellular amount of virus is almost equal to that formed in the cells of infected growing plants. Infection is synchronous and 60–80% of the cells have been infected; about the same amount of virus is produced in 2–3 days in protoplasts as in 10 days in a similar number of cells of an intact growing plant. Protoplasts can also be infected by more than one virus at the same time. Studies such as these might give insight into (1) the effect of one virus on the other in the same host, (2) the mechanism of infection, and (3) nuclear protein replication.

6.7 DNA

Transformation in bacteria by uptake of exogenous DNA is an established fact. However, the information accumulated during the last decade suggests that similar techniques might be used with higher plant cells. Despite some scepticism, these manipulations may yet prove very important. Exogenous DNA can be taken up by higher plant cells/protoplasts [12]. Doy *et al.* [13] claimed to have successfully transformed haploid tomato and *Arabidopsis* callus cultures by *lambda* phage. The fact that these 'transformed' higher plant cells can grow on a medium containing galactose and lactose as the exclusive source of carbon is regarded by these workers to be sufficient proof that phage DNA is functional. These authors used a new term, 'transgenosis', for this type of transformation in a higher plant system. Similar observations have been reported by Johnson *et al.* [14], who incubated sycamore (*Acer pseudoplatanus*) cells in a bacteriophage suspension for 2 days, and then cultured these treated cells on a lactose medium. The control cells died, but the cells incubated with phage continued to grow. They concluded that the lac genes are expressed and this leads to the formation of a *B*-galactosidase which confers the capacity to assimilate lactose.

Most studies have shown that isolated DNA that is taken up by protoplasts is quickly degraded by nucleases. There are special cases for DNA that is intended to be isolated into plant cells such as Agrobacterium plasmids. These are covered in a later section (*see* Section 7.1).

6.8 Non-biological materials

Uptake of non-biological materials such as polystyrene beads may be induced by co-incubating the beads and isolated protoplasts in the presence of PEG. These type of studies have proved useful as a method of determining the maximum size of an object that can be engulfed by endocytosis. However, no direct biological value can be seen to these studies at the present time.

References

[1] Dodds, J. H. and Bengochea, T. (1983) *Outlook on Agric.*, **12**, 16–21.
[2] Bucke, C., Walker, D. A. and Baldry, C. W. (1966) *Biochem. J.*, **101**, 636–41.
[3] Carlson, P. S. (1973) *Proc. Natl. Acad. Sci. (USA)*, **70**, 598–602.
[4] Potrykus, I. (1973) *Z. Pflanzenphysiol.*, **70**, 364–66.
[5] Bonnett, H. T. and Eriksson, T. (1974) *Planta*, **120**, 71–79.
[6] Davey, M. R. and Short, K. C. (1972) *Protoplasma*, **75**, 199–203.
[7] Davey, M. R. and Cocking, E. C. (1972) *Nature (Lond.)*, **239**, 455–56.
[8] Burgoon, A. C. and Bottino, P. J. (1976) *J. Hered.*, **67**, 223–26.
[9] Cocking, E. C. (1966) *Planta*, **68**, 206–14.
[10] Coutts, R. H. A. (1973) in *Protoplasts at fusion de cellules nomatiques vegetales. Colloq. Intern. Centre Natl. Rech. Sci. Paris*, pp.353–65.
[11] Otsuki, Y. and Takebe, I. (1973) *Virology*, **52**, 433–38.
[12] Holl, F. B., Gamborg, O. L., Ohyama, K. and Pelcher, L. (1974) in *Tissue Culture and Plant Science* (ed. H. E. Street), Academic Press, London, pp. 301–27.
[13] Doy, C. H. Gresshoff, P. M. and Rulfe, B. G. (1973) *Proc. Natl. Acad. Sci. (USA)*, **70**, 723–26.
[14] Johnson, C. B., Grierson, D. and Smith, H. (1973) *Nature (New Biol.)*, **244**, 105–106.

7

Genetic engineering

7.1 Introduction

7.1.1 Crown gall tumours

A tumour may be defined as a disorganized growth caused by cellular division that has ceased to be under normal tissue control mechanisms. The great majority of studies have always been on animal tumours. However, tumours are also formed in plants. Plant tumours can be induced by different causal agents—some are genetic, some are viral and others are induced by a bacterium. The last group are known as crown gall tumours and they have been of interest to plant biologists for nearly half a century [1–3]. The name is derived from the fact that a gall tumour forms normally just above soil level, that is at the crown of the plant.

Early studies of gall induction quickly established that the disease was caused as a result of the infection of the plant, at a wound site, with *Agrobacterium tumefaciens*. It soon became of great interest to biologists to determine what infective mechanism of the bacterium was able to cause these gross morphological and biochemical changes in the host plant. It was hypothesized that a 'principle' in the bacterium caused this change and scientists began to hunt for the 'tumour inducing principle' (TIP).

7.1.2 Tumour inducing principle (TIP)

It was quickly established in controlled experimental conditions that the formation of the tumour required:

(1) A wound site.
(2) Inoculation with intact bacteria.
(3) A virulent strain of bacteria.

As was the vogue at that time, scientists then proceeded to lyse open the bacteria and separate them into their cellular components by density gradient centrifugation [4]. They then attempted to infect wounded plants with these sub-cellular extracts. However, none of these extracts were able to initiate the disease and the TIP appeared to be an elusive factor.

In the same period bacteriologists and biochemists working in other areas of investigation were discovering that many bacteria, as well as having the normal bacterial nuclear nucleic acid, also contain many small rings of independent DNA which were named plasmids. In fact for several years some people argued that these fragments of DNA were merely artefacts of nucleic acid isolation.

The interest in plasmids during that era led scientists to study whether *Agrobacterium tumefaciens* contained plasmids and it was striking to note that only the virulent strains of bacteria contained the plasmids. This obviously led to the hypothesis that the infective agent (TIP) is in fact a bacterial plasmid.

Elegant experiments carried out by several research groups have now shown without doubt that the tumour-inducing principle of *Agrobacterium tumefaciens* is a plasmid [5–7]. It was further shown that a portion of the plasmid is actually transferred into the genome of the host plant to effect these changes [8, 9]. This discovery showed that Agrobacterium plasmids were able to introduce new genetic information into a plant genome and this opened the door to studies on plant genetic engineering. The infective plasmid of Agrobacterium is now nicknamed 'nature's genetic engineer'. The use of this plasmid to introduce new genetic information also depended on developments in other areas of molecular biology.

7.1.3 *Principles of recombinant DNA technology*

The genetic information contained within a DNA molecule is dictated by the sequence of the base composition [10]. The basic tools of recombinant DNA technology (genetic engineering) are different enzyme systems that can be used to 'cut and paste' pieces of DNA together. A large number of site-specific restriction enzymes have been identified; these are like molecular scissors that will cut a DNA strand when they encounter a specific sequence. For example, the restriction enzyme Bam I will cut a DNA strand when it encounters the sequence GGATCC. Other enzymes, known as ligases, are used as a molecular sellotape to stick together loose fragments.

By careful choice of enzymes it is now possible to cut open an Agrobacterium plasmid at a single site, thus converting a ring into a strand. A purified or synthetic gene can then be attached to one end. The strand is then resealed into a ring. This transformed Agrobacterium plasmid will now carry this new gene into the plant genome and the idea of genetically engineering plants becomes a reality [11–13].

Let us now look in more detail at the tumours, the plasmids, the role of isolated protoplasts and the agronomic future of these techniques.

7.2 Induced tumours

7.2.1 Experimental induction

To induce tumours experimentally plants are first wounded, then a

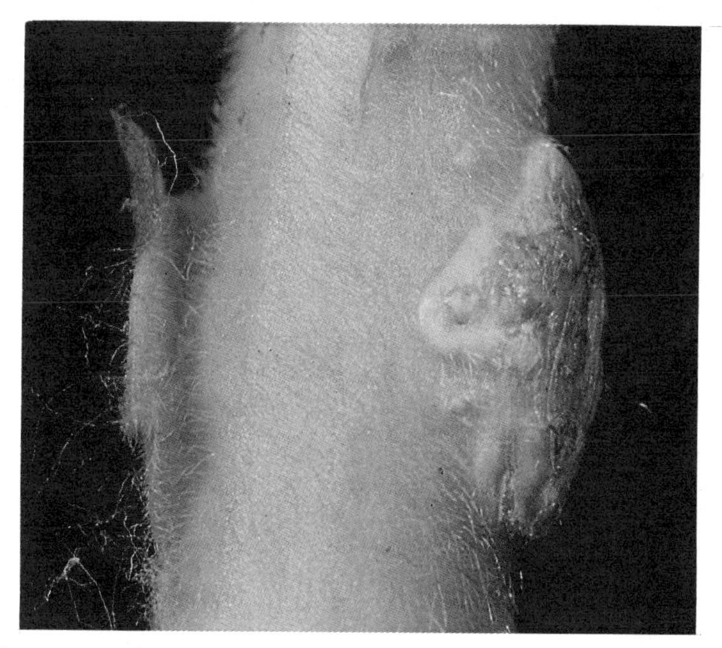

Figure 7.1 Induction of a crown gall on the stem of a sunflower. The stem was wounded and inoculated with *Agrobacterium tumefaciens*.

Figure 7.2 Induction of hairy root infection of the stem of an *in vitro* potato plant. The stem was wounded and inoculated with *Agrobacterium rhizogenes*.

small inoculum of bacteria is applied to the wound surface. This can be done either on intact plants or on *in vitro* plants. *Agrobacterium tumefaciens* will induce a gall at the point of infection (Fig. 7.1). A different organism, *Agrobacterium rhizogenes*, causes the proliferation of hairy roots at the point of infection (Fig. 7.2).

These tumours form because of independent production by the transformed cells of auxins and cytokinins and it thus follows that this tumorous material can be grown on auxin and cytokinin free medium. The tumours also produce a number of unusual amino acids (opines) that are useful molecular markers.

7.3 The infective plasmid

7.3.1 Structure

The last decade has seen a rapid analysis of the mapping of the Agrobacterium plasmid. The basic map of the plasmid is similar to that shown in Fig. 7.3. Without going into details of the molecular biology it is worthwhile to point out some important characteristics of the map. The plasmid contains a segment known as the Ti or (Ri) region. It is this fragment of the plasmid that is translocated and

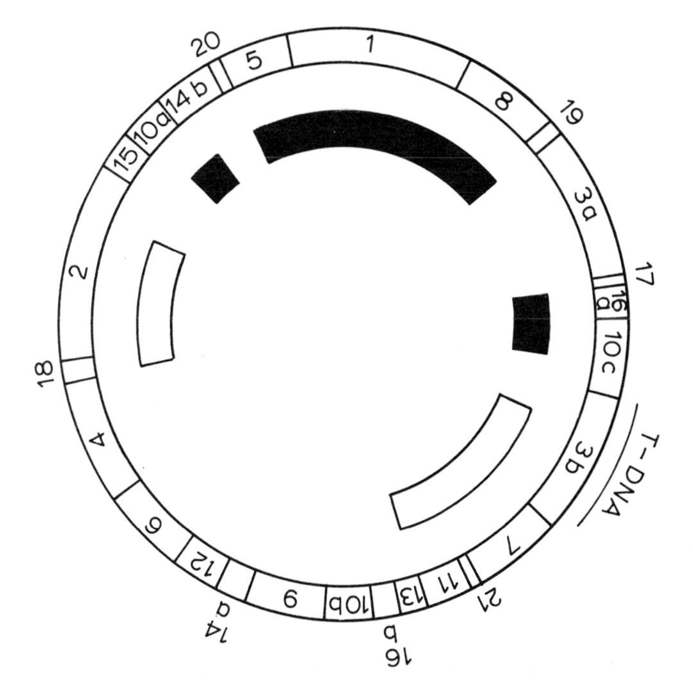

Figure 7.3 Map of *Agrobacterium* plasmid.

incorporated into the plant genome. The plasmid also contains a number of important genes. Some control the formation of auxins and cytokinins whilst others control the synthesis of opines. The plasmid genome fortunately contains a number of useful restriction sites; some of these are unique sites meaning that the ring can be cleared into a single strand for the introduction of new pieces of DNA.

7.3.2 *Modification of the plasmid*

Restriction enzymes can be used to cleave the plasmid at a unique site. New information may then be inserted into the plasmid. This new information may be of a number of types. It may be genes to confer resistance to certain antibiotics; these can be useful as a selectable marker [14]. It may be the insertion of a gene purified from a different organism [15] or it may be a fragment of synthetic DNA, which can act either to produce a novel gene product or as a gene which regulates some other cellular activity [16].

The modified plasmid can then be reinserted to the bacterium. The result is an infective Agrobacterium that can be used as a vector to introduce this foreign information into the plant.

7.4 Transformation of protoplasts

There has been a great deal of interest in transforming protoplasts, especially in those crop species that are not naturally infected by Agrobacter species, for example rice. A number of methods have been used to try to induce uptake of plasmids or bacteria by isolated protoplasts.

7.4.1 *Co-cultivation*

One simple method is simply to add large amounts of plasmid to the surrounding culture medium and allow the protoplast to grow in the media. A small percentage of the protoplast will engulf the plasmid by a process of endocytosis. The frequency of this is relatively low but can be enhanced by heat shock of the protoplasts or the addition of compounds that weaken the plasma membrane. The application of electric impulses (electroporation) is also being studied by a number of people.

7.4.2 *Use of liposomes*

Large numbers of plasmids may be enclosed in a small lipid bag known as a liposome. Protoplasts can then be induced to fuse with the liposomes using methods such as polyethylene glycol as described in Chapter 4. As can be seen diagrammatically in Fig. 7.4 this leads to the introduction of the plasmids into the protoplasts.

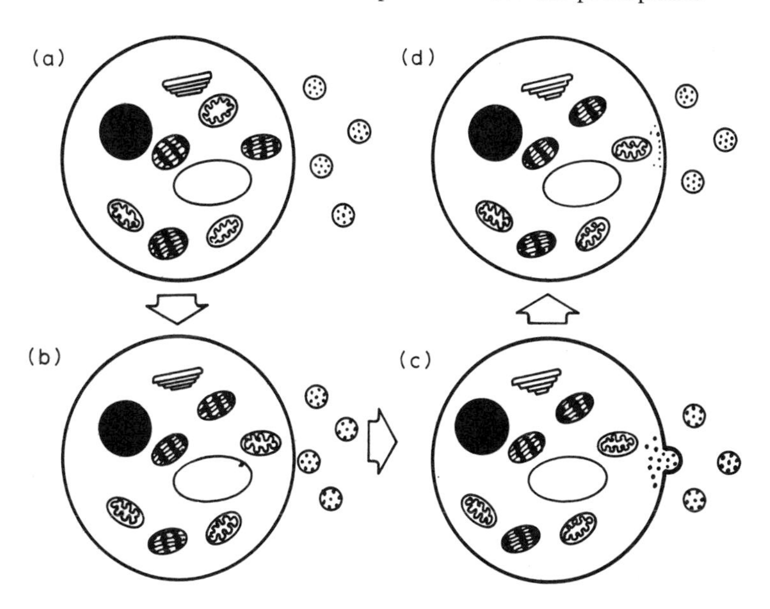

Figure 7.4 Fusion of plasmid-filled liposomes with isolated protoplasts as a method of plasmid uptake.

7.4.3 *Selection of transformed protoplasts*

The protoplasts which incorporate the plasmid into their genome will have certain properties that the non-transformed cells lack. For example, the transformed cells do not require the addition of auxins or cytokinins to the medium to grow. Therefore, if all the protoplasts are planted onto an auxin–cytokinin free medium only the transformed cells will grow. These transformed cells will also produce opines that the normal cells will not. Other selectable characters may have been engineered into the plasmid, i.e. antibiotic resistance.

The technology is now developing rapidly to allow the use of Agrobacterium plasmids and protoplasts to play a role in plant improvement. The major problem facing molecular biologists is which genes may be of use to improve crop plants.

7.5 Which genes to engineer?

7.5.1 Model genes

For these types of studies legume storage proteins are being used as a model system. As well as being extremely important nutritional proteins they have the advantage of being produced site specifically in large quantitites at a specific developmental stage. These factors have allowed the easy isolation of large quantities of messenger RNA (mRNA). A viral enzyme called reverse transcriptase may be used to produce a copy DNA (cDNA) from an mRNA. In other words the cDNA is in essence a copy of the legume storage protein gene. It has been recently demonstrated that a cDNA of a protein storage gene can be inserted into petunia plants using a Ti plasmid vector. The seeds of the petunia plant then produce bean proteins [17].

7.5.2 Synthetic genes

In the case of storage protein genes the previous technique is a most valuable one to produce copies of genes. However, this type of technique can be applied only in a very limited number of cases. To isolate, purify and clone genes or gene products that in the plant are only produced in low concentration is still beyond the scope of molecular biologists. Therefore, in the short term an alternative may

```
SP44 (1) →→→

AATTCGGGACGATCACGATCCATCCATTTCTTAAGAAATGGATGACGATCCATCCATTTCTTAAGAAATGGATG
    GCCCTGCTAGTGCTAGGTAGGTAAAGAATTCTTTACCTACTGCTAGGTAGGTAAAGAATTCTTTACCTAC

ACGATCCATCCATTTCTTCCCG
TGCTAGGTAGGTAAAGAAGGGCTTAA

(a)                                 ←←← SP44 (2)

SP44 (1)

    GlyThrIleThrIleHisProPheLeuLysLysTrpMetThrIleHisProPheLeuLysLysTrpMet
(b) ThrIleHisProPheLeuPro
```

Figure 7.5 Structure of a synthetic gene. The DNA base sequence (a) can be translated to produce an amino acid sequence (b) that is enriched in essential amino acids.

be the use of synthetic genes.

Strings of DNA that are effectively synthetic genes may be synthetized chemically. Custom-made machines (gene-machines) can now be programmed to do this.

The synthetic genes can be at least 500 bases in length. Fig. 7.5 shows a synthetic gene sequence. This gene, when translated, produces a synthetic protein which is rich in essential amino acids. By coupling together gene synthesis, plasmid technology and protoplast culture this type of gene may be inserted into economically important crop plants [18]. This possibility coupled with conventional plant breeding methods could well open the door to a second green revolution [19].

References

[1] Braun, A. C. (1943) *Am. J. Bot.*, **30,** 674–77.
[2] Braun, A. C. (1978) *Biochem. Biophys. Acta*, **516,** 167–91.
[3] Braun, A. C. (1982) in *Molecular Biology of Plant Tumors* (ed. G. Kahl and J. Schell), Academic Press, New York, pp. 155–210.
[4] Quail, H. (1980) *Plant Cell Fractionation.* Plenum Press, New York.
[5] Van Larebeke, N., Engler, G., Holsters, M., Van der Elsaker, S., Zaenen, I., Schilperoott, R. A. and Schell, J. (1974) *Nature*, **252,** 169–70.
[6] Zaenen, I., Van Larebeke, N., Teuchy, H., Van Montagne, M. and Schell, J. (1974) *J. Molec. Biol.*, **86,** 109–27.
[7] Watson, B., Currier, T. C., Gordon, M. P., Chilton, M. D. and Nester, E. W. (1975) *J. Bacteriol.*, **123,** 255–64.
[8] Zambryski, P., Joos, H., Genetello, C., Leemans, J., Van Montague, M. and Schell, J. (1983) *EMBO J.*, **2,** 2143–50.
[9] Herrera-Estrella, L., De Block, M., Zambryski, P., Van Montague, M. and Schell, J. (1985) in *Plant Genetic Engineering* (ed. J. H. Dodds), Cambridge University Press, Cambridge, pp.61–93.
[10] Watson, J. D. (1965) *Molecular Biology of The Gene.* W. A. Benjamin, New York.
[11] Dodds, J. H. (ed.) (1985) *Plant Genetic Engineering*, Cambridge University Press, Cambridge, p. 312.
[12] Bengochea, T. and Dodds, J. H. (1983) *Outlook on Agric.*, **12,** 16–20.
[13] Lea, P. and Stewart, G.R. (1984) *Genetic Manipulation of Plants and its Application to Agriculture.* Oxford University Press, Oxford.
[14] De Block, M., Herrera-Estrella, L., Van Montague, M., Schell, J. and Zambryski, P. (1984) *EMBO J.*, **3,** 1681–89.
[15] Herrera-Estrella, L., Depicker, A., Van Montague, M. and Schell, J. (1983) *Nature (Lond.)*, **303,** 209–13.

[16] Jaynes, J.M. and Dodds, J.H. (1986) *Plant Cell Rep.* (in press).
[17] Croy, R. R. D. and Gatehouse, J. A. (1985) in *Plant Genetic Engineering* (ed. J.H. Dodds), Cambridge University Press, Cambridge, pp. 143–268.
[18] Jaynes, J. M., Espinoza, N. O. and Dodds, J. H. (1986) *Sci. Am.* (in press).
[19] *Business Week*. The 2nd Green Revolution. August 25th (1980).

Index